T0133565

Cloud Computing in Remote Sensing

Cloud Computing in Remote Sensing

Lizhe Wang
Jining Yan
Yan Ma

CRC Press is an imprint of the
Taylor & Francis Group, an **informa** business

CRC Press
Taylor & Francis Group
6000 Broken Sound Parkway NW, Suite 300
Boca Raton, FL 33487-2742

Printed on acid-free paper

International Standard Book Number-13: 978-1-138-59456-2 (Hardback)

Visit the Taylor & Francis Web site at
http://www.taylorandfrancis.com

and the CRC Press Web site at
http://www.crcpress.com

Contents

Preface

With the remarkable advances in latest-generation high-resolution Earth Observation (EO), the amount of remotely sensed (RS) data has been accumulated to exabyte-scale and been increasing in petabytes every year. Concerning the present and upcoming high-resolution mission, much higher spatial, spectral or temporal resolution sensors probably give rise to higher dimensionality of data. Meanwhile, planetary-scale applications in earth science or environmental studies may inevitably aggravate the complexity of data and its processing, due to a variety of multi-sensor and multi-resolution large datasets introduced. Therefore, RS data have been gradually accepted as "Big RS Data", not merely for the extreme data volume, but also the complexity of data.

The explosive growth of RS big data has even been revolutionizing the way RS data are managed and analyzed. The significant data volume is far beyond the storage capability traditional data centers could satisfy, since it is remarkably complicated and expensive to be frequently upgraded and scaled-out to guarantee a growth rate of petabytes per year. Great challenges have also been introduced in the management of the multi-sensor, multi-spectral, multi-resolution and multi-temporal featured large datasets that might be in various formats or distributed across data centers. Likewise, exceptional computational challenges have been routinely introduced by the applications that extend their models to continental or global scale, and even use temporal data for long time-serial analysis. The case could become even worse especially because of the spurt of interest in on-the-fly processing that demands tremendous data processing in an extremely short time. Even though the high-performance computing(HPC) resources could mainly facilitate large-scale RS data processing, they are almost out of the reach of many researchers, due to the burden of considerable technical expertise and effort. Accordingly, most existing models hardly accommodate large datasets at global scale, as they are commonly applied on a limited number of datasets within some small area.

Why Cloud computing in remote sensing? Taking advantage of elasticity and high-level of transparency, Cloud computing offers a revolutionary paradigm that resources are accommodated as ubiquitous services on-demand on a pay-as-use basis along with both flexibility and scalability. Incorporation of Cloud computing to facilitate large-scale scientific applications turns out to be a widespread yet easy-to-use approach to the tremendous computational and storage challenges. In light of this, our book focuses great efforts on Cloud-enabled integrated platforms and techniques to empower petabyte-scale RS big data managing, planetary-scale on-the-fly analyzing, together with easy-to-use RS data and processing services on-demand.

Accordingly, this book begins with a cloud-based Petabyte-scale RS data integration and management infrastructure. Based on an OpenStack-enabled Cloud infrastructure with flexible resources provisioning, a distributed RS data integration system is built to easily integrate petabyte-scale multi-sensor, multi-resolution RS data in various formats or metadata standards across data centers. By virtue of the logical segmentation indexing (LSI) global data organizing and spatial indexing model as well as the distributed No-SQL database, an optimized object oriented data technology (OODT) based RS data management system is put forward to serve RS data on-demand. Then, the book is followed by high-performance remote sensing clouds for the data center named pipsCloud. It incorporates the Cloud computing paradigm with cluster-based HPC systems for a quality of service (QoS) optimized Cloud so as to address the challenges from a system architecture point of view. Wherein, bare-metal (BM) provisioning of the HPC cluster is proposed for less performance penalty; a data access pattern aware data layout strategy is employed for better data locality and finally optimized parallel data I/O; dynamic directed acyclic graph (DAG) scheduling is used for large-scale workflows. Likewise, benefitting from an optimal distributed workflow and resource scheduling strategy on a basis of minimal data transferring, Cloud-enabled RS data producing infrastructure and services are delivered for planetary-scale RS data analyzing across data centers. In addition, the application-oriented generic parallel programming technique is also discussed for easy cloud-empowered large-scale RS data processing. Last but not the least, the RS information analysis, knowledge discovering, RS data and processing service models are all discussed here.

This book is greatly supported by the National Natural Science Foundation of China (No. U1711266), and it would appeal to remote sensing researchers becoming aware of the great challenges lying in most earth sciences as well as Cloud-empowered infrastructures and techniques available. Meanwhile, the computer engineers and scientists from other domains with similar issues could also be benefitted or inspired.

Chapter 1

Remote Sensing and Cloud Computing

1.1 Remote Sensing

Remote sensing is a new comprehensive detection technology in the 1960s. Remote sensing technology, characterized by digital imaging, is an important symbol to measure a country's scientific and technological development level and comprehensive strength.

1.1.1 Remote sensing definition

Remote sensing is generally defined as the technology of measuring the characteristics of an object or surface from a distance [1, 2]. It is the acquisition of information about an object or phenomenon without making physical contact with the object and thus stands in contrast to on-site observation. This generally refers to the use of sensors or remote sensors to detect the electromagnetic radiation and reflection characteristics of objects. The remote sensing technology acquires electromagnetic wave information (such as electric field, magnetic field, electromagnetic wave, seismic wave, etc.) that is reflected, radiated or scattered, and performs scientific techniques of extraction, measurement, processing, analysis and application.

At present, the term "remote sensing" generally refers to the detection and classification of objects on the earth, including the surface, the atmosphere and the oceans, based on transmitted signals (such as electromagnetic radiation), using sensor technology on satellites or aircraft [3]. It is also divided into "active" remote sensing and "passive" remote sensing.

1.1.2 Remote sensing big data

With the rapid development of remote sensing technology, our ability to obtain remote sensing data has been improved to an unprecedented level. We have entered an era of big data. Remote sensing data clear showing the characteristics of Big Data.

(1) Big volume

According to incomplete statistics, the total amount of data archived by the Earth Observing System Data and Information System (EOSDIS) reached 23.8 petabytes (PBs) around the year 2017, and kept 15.3 TB/day average archive growth speed. Up until October 2018, the archived data volume of the China National Satellite Meteorological Center (NSMC) reached 4.126 PBs, and the China Center for Resources Satellite Data and Application (CCRSDA) archived more than 16 PBs of remote sensing images until the end of 2016.

(2) Big variety

According to the 2017 state of the satellite industry report, there were 277 earth observation satellites in orbit by the end of 2016. These satellites all carried more than one earth observation sensor, and they can collect earth surface information continuously, day and night. That is to say, at least 277 kinds of earth observation data will be continuously transmitted to the ground receiving station. In addition, since Landsat-1 first started to deliver volumes of pixels in 1972, there have been more than 500 earth observation satellites launched into space, with archived remote sensing data of more than 1000 types.

(3) Big velocity

With the development of multi-satellite coordination and satellite constellation combination observation technologies, the satellite revisit periods gradually transition from month to day, hour or even minute. For example, the Jilin-1 satellite constellation consisting of 60 satellites will have a revisit cycle of 20 minutes by the end of 2020. In addition, the remote sensing data received by each data center arrives continuously at an ever-faster code rate. For example, the amount of data received from a GF-2 satellite PSM1 sensor is approximately 1.5 TB per day. It is preferable to ingest and archive the newly received data in order to provide users with the latest data retrieval and distribution service.

(4) Big value but small density

With spatial resolution and spectral resolution increases, more and finer ground information features could be captured by satellite sensors. Using remote sensing images, we can investigate Land-Use and Land-Cover Change (LUCC), monitor vegetation growth, discover ground military targets, etc. However, in order to obtain such valuable information, we have to process massive remote sensing images, just like picking up gold from the sand.

(5)Heterogeneous

Due to various satellite orbit parameters and the specifications of different sensors, the storage formats, projections, spatial resolutions, and revisit periods of the archived data are vastly different, and these differences have resulted in great difficulties for data collection. For example, Landsat 8 collects images of the Earth with a 16-day repeat cycle, referenced to Worldwide Reference System-2. The spatial resolution of the Operational Land Imager (OLI) sensor onboard the Landsat 8 satellite is about 30 meters; its collected images are stored in GeoTIFF format, with Hierarchical Data Format Earth Observation System (HDF-EOS) metadata. The Moderate Resolution Imaging Spectroradiometer (MODIS) instruments capture data in 36 spectral bands ranging in wavelength from 0.4 µm to 14.4 µm and at varying spatial resolutions (2 bands at 250 m, 5 bands at 500 m, and 29 bands at 1 km). Most of the MODIS data is available in the HDF-EOS format, and it is updated every 1 to 2 days. The Charge Coupled Device (CCD) sensor, which is carried by the Huan Jing (HJ)-1 mini satellite constellation, has an image swath of about 360 km, with blue, green, red, and NIR bands, 30m ground pixel resolution, and a 4-day revisit period. Its collected images are stored in GeoTIFF format, and their customized metadata is in eXtensible Markup Language (XML) format.

(6) Offsite storage

Generally speaking, different types of remote sensing data sources are stored in different satellite data centers, such as wind and cloud satellite data stored in meteorological satellite centers, and marine remote sensing data stored in marine satellite centers. In order to maximize the use of these earth observation and earth exploration data to serve us, we need to collect the offsite stored big earth data for unified management.

1.1.3 Applications of remote sensing big data

Now, remote sensing big data are attracting more and more attention from government projects and commercial applications to academic fields. It has been widely used in agriculture, disaster prevention and reduction, environmental monitoring, public safety and urban planning and other major macro application decisions, because of its advantages of obtaining large-scale and multi-angle geographic information [4].

In March 2012, the United States government proposed the "Big Data" Initiative. It could be the first government project on big data that focuses on improving our ability to extract knowledge from large and complex collections of digital data. For remote sensing big data, one of the most important US government projects is the Earth Observing System Data and Information System (EOSDIS). It provides end-to-end capabilities for managing NASA's Earth science data from various sources. In Europe, the "Big Data from Space" conference was organized by the European Space Agency in 2017. It is to stimulate interactions and bring together researchers, engineers, users, infrastructure and service providers interested in exploiting big data from Space. In October 2017, the group on Earth Observations (GEO), the largest intergovernment multi-lateral cooperation organization, promotes the development of big data. Besides, the National GEOSS Data Sharing Platform of China is also delivered in the GEO Week 2017.

In the field of commercial applications, Google Earth could be one of the examples of success of remote sensing big data. Many remote sensing applications such as target detection, land-cover, smart city, etc. can be developed easily based on Google Earth. With the Digital Globes Geospatial Big Data platform (GBDX) ecosystem, the Digital Global company (Longmont, CO, USA) is building footprints quickly by leveraging machine learning in combination with Digital Globes cloud-based 100 petabyte imagery library. Other large companies such as Microsoft (Redmond, WA, USA) and Baidu (Beijing, China) are all developing their electronic maps that are supported with remote sensing big data and street views big data. The commercial applications for big data are changing peoples' lives.

In academic fields, remote sensing big data is also one of the most popular topics. Many top journals have launched their special issues about remote sensing big data. IEEE JSTARS launched a special issue on Big Data in Remote Sensing in 2015. The Journal of Applied Remote Sensing launched a special issue on Management and Analytics of Remotely Sensed Big Data in 2015. IEEE Geoscienceand Remote Sensing Magazine launched a special issue on Big Data from Space in 2016. GeoInformatica of Springer launched a special issue on Big Spatial and Spatiotemporal Data Management and Analytics in 2016. Environmental Remote Sensing launched a special issue on Big Remotely Sensed Data: Tools, Applications and Experiences in 2017. Remote Sensing MDPI is calling for papers on special issues in Advanced Machine Learning and Big Data Analytics in Remote Sensing for Natural Hazards Management, SAR in the Big Data Era and Analysis of Big Data in Remote Sensing in 2018. The International Journal of Digital Earth is calling for papers for the special issue on Social Sensing and Big Data Computing for Disaster Management in 2018.

No matter whether it is government projects, commercial applications or academic research, when characterizing big data, it is popular to refer to the 3Vs, i.e., remarkable growth in Volume, Velocity and Variety of data [5]. For remote sensing big data, they could be more concretely extended to characteristics of multi-source, multi-scale, high-dimensional, dynamic-state, isomer, and nonlinear characteristics. It is important for us to consider these more concrete and particular characteristics of remote sensing big data when using remote sensing to extract information and understand geo-processes. It is both an opportunity and a challenge for remote sensing communities. With these issues in mind, this book presents the current, state-of-the-art theoretical, methodological, and applicational research on remote sensing big data.

1.1.4 Challenges of remote sensing big data

Remote sensing big data has posed great challenges for data integration and processing.

1.1.4.1 Data integration challenges

(1) A unified data standard is urgently needed for heterogeneous data integration: Data integration requires uniform data standards, including metadata standards and image standards. However, due to the massive, multi-source and heterogeneous characteristics of big remote sensing data, it is challenging to make a unified data integration standard.

(2) Network transmission efficiency and data security are not guaranteed during data integration: During the process of data integration, network transmission efficiency is the key to obtain data successfully. Furthermore, data is easily attacked and intercepted during transmission, and data security is also critical.

(3) The synchronization problem is unavoidable after data integration: The data storage and management system of different data centers has a high degree of autonomy, and has the ability to add, update, and delete autonomously for stored data. Therefore, data collection and integration needs to address data consistency issues between the original data storage management system and the integrated system.

1.1.4.2 Data processing challenges

(1) Efficient application-specific data management and storage strategies: How to design efficient application-specific data management and storage strategies and provide applications a unified interface for easy access from distributed collections of big earth data are a key challenge. They can improve the performance of data-intensive big earth applications in the following two aspects: a) application-specific data management and storage strategies take the application's data access patterns into consideration, so the computing system has the ability to realize high throughput; b) in a computing system, data transmission is a time-consuming task due to the limited network bandwidth. A proper and large data structure significantly decreases the data transmission time, especially when the volume of communication is extremely large.

(2) Proper earth data selection to address real-world earth applications: Big remote sensing data includes different kinds of data with different data formats. Normally, those different data formats from different data sources are combined to fulfill big earth applications. Hence, how to select and combine different formats of data to address real-world earth applications is a key challenge for big earth data.

(3) Resources and tasks scheduling: The processing of these applications becomes extremely difficult because of the dependency among a large collection of tasks which give rise to ordering constraints. The succeeding tasks have to wait for the output data of preceding tasks to be available. The optimized scheduling of these bunches of tasks is critical to achieve higher performance.

1.2 Cloud Computing

Cloud computing has emerged as a hot topic among IT industries, academics and individual users due to its abilities of offering flexible dynamic IT infrastructures, QoS guaranteed computing environments and configurable software services [6, 7, 8]. According to the definition of the National Institute of Standards and Technology (NIST), cloud computing is regarded as "*a model for enabling ubiquitous, convenient, on-demand network access to a shared pool of configurable computing resources (e.g., networks, servers, storage, applications, and services) that can be rapidly provisioned and released with minimal management effort or service provider interaction*" [9]. Cloud computing has five essential characteristics: on-demand self-service, broad network access, resource pooling, rapid elasticity and measured service. Based on the definition of NIST, cloud computing has three service models and four deployment models.

1.2.1 Cloud service models

Cloud computing employs a service-driven model; services of clouds can be grouped into three categories: software as a service (SaaS), platform as a service (PaaS) and infrastructure as a service (IaaS).

(1) Infrastructure as a Service: IaaS adopts virtualization technologies to provide consumers with on-demand provisioning of infrastructural resources (e.g. networks, storages, virtual servers etc,.) on a pay-as-you-go basis. IaaS helps consumers avoid the expense and complexity of buying and managing physical servers and other datacenter infrastructure. Consumers are able to quickly scale up and down infrastructural resources on demand and only pay for what they use. The cloud owner who offers IaaS is called an IaaS provider. Companies providing IaaS include Amazon, Google, Rackspace, Microsoft, etc.

(2) Platform as a Service: Cloud providers manage and deliver a broad collection of middleware services (including development tools, libraries and

database management systems, etc.). Consumers adopt PaaS to create and deploy applications without considering the expense and complexity of buying and managing software licenses, the underlying application infrastructure and middleware or the development tools and other resources.

(3) Software as a Service: SaaS is a model for the distribution of software where customers access software over the Internet on a pay-as-you-go basis. Normally, consumers access software using a thin client via a web browser.

1.2.2 Cloud deployment models

Cloud deployment refers to a cloud that is designed to provide specific services based on demands of users. A deployment model may embrace diversified parameters such as storage size, accessibility and proprietorship, etc. There are four common cloud deployment models that differ significantly: Public Clouds, Community Clouds, Private Clouds and Hybrid Clouds.

(1) Public Cloud: Public clouds refer to the cloud that service providers offer their resources as services to the general public or a large industry group. In order to ensure the quality of cloud services, Service Level Agreements (SLAs) are adopted to specify a number of requirements between a cloud services provider and a cloud services consumer. However, public clouds lack fine-grained control over data, network and security settings.

(2) Private Cloud: Private clouds are designed for exclusive use by a particular institution, organization or enterprise. Comparing with public clouds, private clouds offer the highest degree of control over performance, reliability and substantial security for services (applications, storage, and other resources) provided by service providers.

(3) Community Cloud: Community clouds are built and operated specifically for a specific group that have similar cloud requirements (security, compliance, jurisdiction, etc.).

(4) Hybrid Cloud: Hybrid clouds are a combination of two or more clouds (public, private or community) to offer the benefits of multiple deployment models. Hybrid clouds offer more flexibility than both public and private clouds.

1.2.3 Security in the Cloud

As is shown in Figure 1.1, the data security lifecycle includes five stages: create, store, use, archive, and destruct. In the create stage, data is created by client or server in the cloud. The Store stage means generated data or uploaded data are stored in the cloud across a number of machines. During the Use stage, data is searched and extracted from the cloud environment. Rarely used data is archived in an other place in the cloud. In the Destroy stage, users have the ability to delete data with certain permissions.

Based on the data security lifecycle, three apects need to be taken into consideration when talking about the security in a cloud: confidentiality,

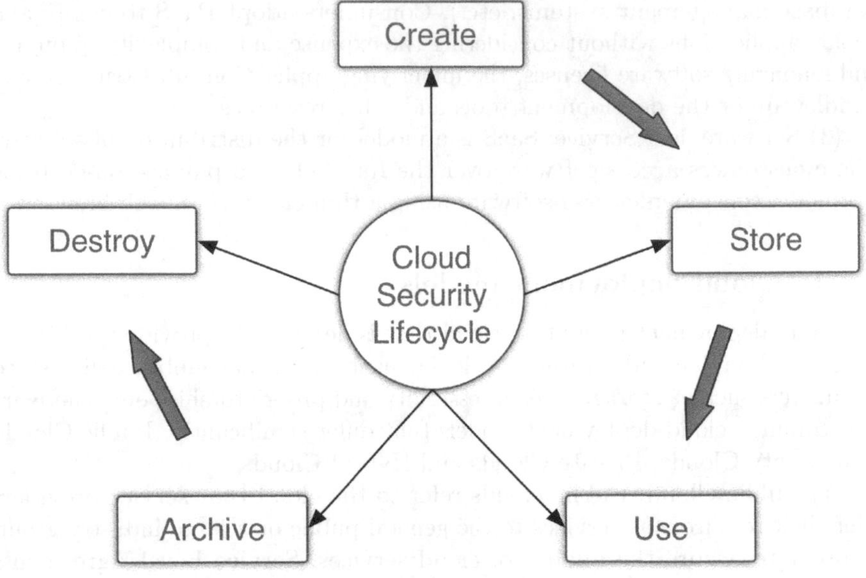

FIGURE 1.1: Data security lifecycle.

integrity, and availability. Confidentiality means the valuable data in the cloud can only be accessed by authorized parties or systems. With the incremental number of parties or systems in the cloud, there is an increase in the number of points of access, resulting in huge threats for data stored or archived in the cloud. In the cloud, resources and services are provided in a pay-as-you-go fashion; integrity protects resources and services paid by consumers from unauthorized deletion, modification or fabrication. Availability is the metric to describe the ability of a cloud to provide resources and services for consumers.

1.2.4 Open-source Cloud frameworks

Currently, there are several representative open-source softwares developed to deploy cloud environments on a number of commodity machines. Figure 1.2 is the web search volume index of three popular cloud computing frameworks in terms of OpenStack, Apache CloudStack, and OpenNebula in Google Trends during the period from January 1, 2008 to October 24, 2018.

1.2.4.1 OpenStack

OpenStack is an open source software platform combining a number of open source tools to build and manage private and public clouds. OpenStack is mostly deployed as infrastructure-as-a-service (IaaS), whereby pooled virtual resoures (compute, storage, and networking) are made available to customers.

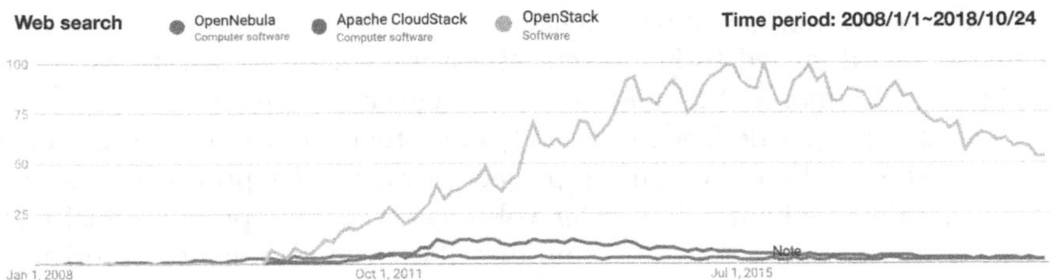

FIGURE 1.2: Web search volume index of three popular open-source cloud computing frameworks in Google Trends. (https://trends.google.com/trends/explore?date=2008-01-01%202018-10-24&q=%2Fm%2F0g58249,%2Fm%2F0cnx 0mm,%2Fm%2F0cm87w_)

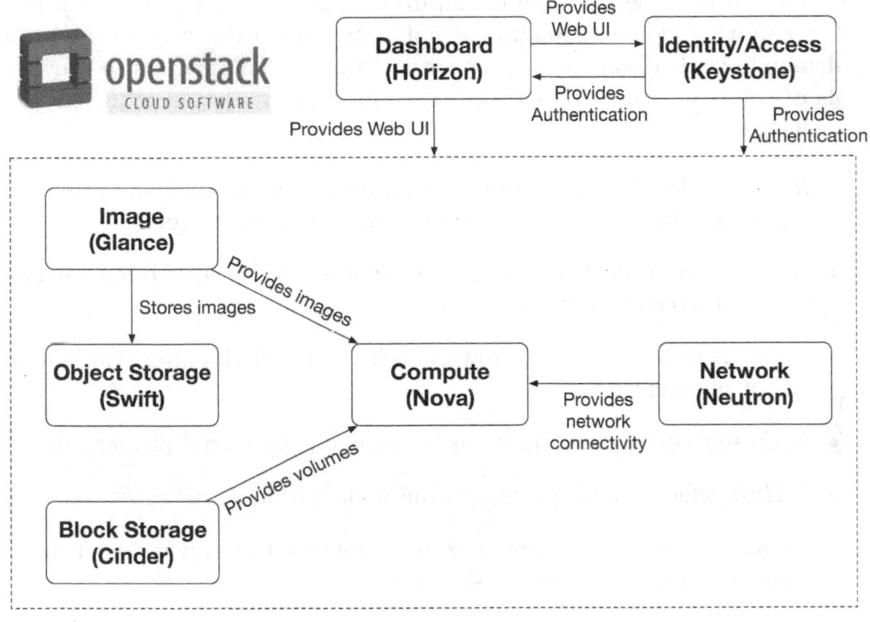

FIGURE 1.3: OpenStack components.

As is shown in Figure 1.3, OpenStack consists senven components. *Nova* is designed to provide on-demand access to computing resources by provisioning and managing computing instances (aka virtual machines). *Neutron* is the project focused on delivering networking-as-a-service (NaaS) in virtual computing environments. In OpenStack, the neutron is designed to provide networking support for the cloud environment. *Swift* is a project that aims to provide a highly available, distributed, eventually consistent object/blob store in OpenStack. Data are stored as binary objects on the server operating system's underlying file system efficiently, safely, and cheaply. *Glance* is the image service designed to help users discover, register, and retrieve virtual machine (VM) images. Virtual machine images are normally stored in *Swift*. *Cinder* is the project designed to provide persistent block-level storage devices for use with OpenStack computing project nova. *Cinder* provisions and manages block devices known as Cinder volumes. *Keystone* provides API client authentication, service discovery, and distributed multi-tenant authorization. *Horizon* is the OpenStack dashboard project aimed to provide a web based user interface to other OpenStack services including Nova, Swift, Keystone, etc.

1.2.4.2 Apache CloudStack

Apache CloudStack [10] is an open source software that manages pools of resources (storage, network, and computer, etc.) to deploy public or private IaaS computing clouds. CloudStack not only provides a way to set up an on-demand elastic cloud computing service, but also allow users to provision pools of resources. Resources within the cloud are managed with following terminologies.

- *Regions* refers to a collection of a number of geographically proximate zones that are managed by one or more management servers.
- A *zone* is equivalent to a single datacenter, and consists of one or more pods and secondary storage.
- A *Pod* is usually a rack, or row of racks that includes layer-2 switch and one or more clusters.
- A *Cluster* contains a number of homogenous hosts and primary storage.
- A *Host* refers to a single computing node within a cluster.
- *Primary storage* is a storage resource provided to a single cluster for the actual running of instance disk images.
- *Secondary storage* is a zone-wide resource that stores disk templates, ISO images, and snapshots.

1.2.4.3 OpenNebula

OpenNebula [11] is an open source cloud solution that manages heterogeneous distributed data centre infrastructures to enable private, public and hybrid IaaS clouds. As is shown in Figure 1.4, an OpenNebula system contains the following basic components.

- *Front-end* executes the OpenNebula services.

- *Hosts* are responsible for creating VMs accroding to the resources requested by users.
- *Datastores* holds the base images of the VMs.
- *Networks* are adopted to support basic services such as interconnection of the storage servers and OpenNebula control operations, and VLANs for the VMs.

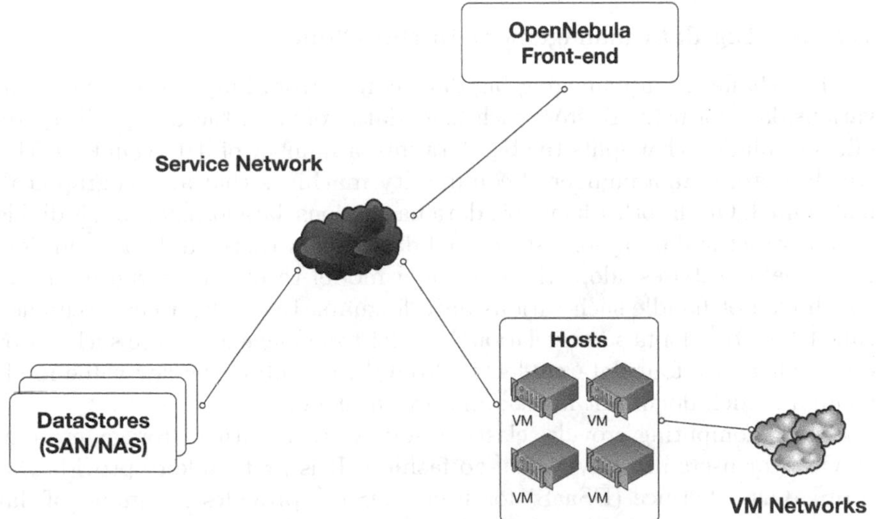

FIGURE 1.4: Cloud architecture of OpenNebula.

Two main features provided by OpenNebula are Data Center Virtualization and Cloud Infrastructure.

- *Data Center Virtualization Management*: OpenNebula directly integrates with hypervisors (like KVM, Xen or VMware ESX) and has complete control over virtual and physical resources. In this way, OpenNebula provides the ability to manage data center virtualization, consolidate servers, and integrate existing IT assets for computing, storage, and networking, and provides advanced features for capacity management, resource optimization, high availability and business continuity.
- *Cloud Management*: Based on an existing infrastructure management solution, OpenNebula builds a multi-tenant, cloud-like provisioning layer to enable users provisioning, elasticity and multi-tenancy cloud features.

1.2.5 Big data in the Cloud

The notion of big data is adopted to refer to large volumes of data that cannot be captured, managed, and processed by general computers within an acceptable scope. Volume, velocity and variety are the mostly used characteristics to describe big data. Volume refers to the amount of big data, velocity means the rate at which the data is generated, and variety represents different data types. Big data poses new challenges in managing and processing such large volumes of data.

1.2.5.1 Big data management in the Cloud

The challenge in managing big data comes from huge data volumes and various data formats. To store such huge data volumes, the Google file system offers a solution that splits the big data into a number of data chunks, each of which is stored in a number of commodity machines that are geographically distributed. On the other hand, big data has various data formats, and is divided into unstructured data, semi-structured data, and structured data. Traditional relational databases adopt the relational model to manage structured data, which cannot handle such various data formats. Recently, a new technology called NoSQL adopts a non-relational model to manage and store such various data. There are four data models in NoSQL, including the key-value model, column model, document model, and graph model.

Cloud computing provides elastic resources (computing, storage, etc.) and services for users in a pay-as-you-go fashion. It is preferable to provide Data Storage as a Service (DSaaS) for users. DSaaS provides a number of data management system (such as RDBMS, NoSQL, and NewSQL) services to help users organize, manage and store big data in the cloud.

1.2.5.2 Big data analytics in the Cloud

Two big data processing framworks including MapReduce and Spark are introduced to process big data stored in the cloud.

(1) MapReduce

MapReduce [12] was proposed by Google in 2004, and has emerged as the de facto paradigm for large scale data processing across a number of commodity machines. The procedures in MapReduce can be divided into the Map stage and the Reduce stage. In the Map stage, the data is divided into a number of data splits; each of them is processed by a `mapper` task. The `mapper` task reads data from its corresponding data split and generates a set of intermediate key/value pairs. These intermediate key/value pairs are then sent to different partitions; each partition is processed by a `reducer` task in the Reduce stage, and generates the final results. To enhance the performance of MapReduce applications, an optional procedure called `Combiner` is often used to gather key/value pairs with the same key. The number of intermediate key/value pairs processed by `reducer` tasks can be significantly alleviated after the `Combiner` is executed, which results in efficiency improvement.

Apache Hadoop [13] is a framework developed to process large scale datasets across distributed clusters of computers using simple programming models. It is an open source implementation of the Google File System and the MapReduce paradigm of Google. Hadoop is composed of two main components: the Hadoop Distributed File System (HDFS) [14] and the MapReduce paradigm. HDFS is the persistent underlying distributed file system of Hadoop, and is capable of storing terabytes and petabytes of data. HDFS has two components, the Namenode managing the file system metadata and the Datanode storing the actual data. The data stored in HDFS is divided into a number of fixed-size (e.g. 128 MB) data chunks. Each data chunk as well as its replicas (normally three replicas) are stored in a number of datanodes in turn. The data stored in HDFS can be parallel processed by the MapReduce paradigm.

As one of the top-level Apache projects, Hadoop has formed a Hadoop-based ecosystem [15], which becomes the cornerstone technology of many big data and cloud applications [16].

(2) Spark

Spark was developed by UC Berkeley AMPLab to provide a unified analytics engine for large-scale data processing, and has become another big data processing mainstream. Figure 1.5 illustrates the architecture of Spark. Once a spark program is submitted to the cluster, an object called *SparkContext* is created to *Cluster Manager* (currently Spark only supports three cluster managers: Apache Mesos, Hadoop YARN, and Kubernetes). Cluster Manager is responsible for allocating resources across applications. Once connected, SparkContext requires a number of *Executors* on nodes in the cluster. Executors on each node are responsible for running tasks and keeping data in memory or disk storage across them. After the executors are allocated, SparkContext sends application code to the executors. At last, the SparkContext sends tasks to the executors to run.

The ability of data to persist in memory in a fault-tolerance manner makes Spark more suitable for numerous data analytics applications, especially iterative jobs. This ability is realized with Resilient Distributed Dataset (RDD) [17], a data abstraction for big data analytics. Substantially, RDD is a coarse-grained deterministic immutable data structure with lineage-based fault-tolerance. RDD has two aspects: data modeling and job scheduling.

Data modeling: RDD is an abstraction of a read-only distributed dataset. An RDD is a read-only collection of objects partitioned across a number of nodes that can be rebuilt. RDDs achieve fault tolerance through a notion of lineage. A new RDD can be generated once coarse-grained RDD transformations are applied to all the data items in the primary RDD. Each RDD object contains a pointer to its parent and information about how the parent was transformed and forms a long lineage graphs. With the help of lineage, if a partition of an RDD is lost, the RDD is able to rebuild that partition with sufficient information about how it was derived from other RDDs.

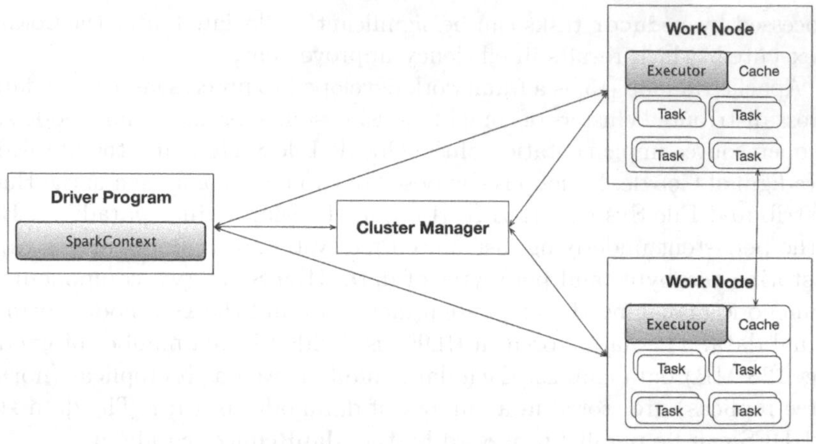

FIGURE 1.5: The Apache Spark architecture.

Job scheduling: In Spark, jobs are organized into a Directed Acyclic Graph (DAG) of stages. Note that an RDD uses lazy materialization, which means that an RDD is not computed unless it is used in an action. When an action is executed on an RDD, the scheduler checks the RDD's lineage in order to build a DAG of jobs for execution. Spark first organizes the jobs into a DAG of different stages, each of which contains a sequence of jobs. The boundaries of different stages are the operations with shuffle. During each stage, a task is formed by a sequence of jobs on a partition, and is performed by executors.

Currently, as one of the core components, Spark is deeply integrated with the Berkeley Data Analytics Stack (BDAS) [18], which aims to develop an open source, next-generation data analytics stack under development at the UC Berkeley AMPLab. In BDAS, Spark SQL, Spark Streaming, MLlib and GraphX are respectively built on top of Spark for SQL-based manipulation, stream processing, machine learning and graph processing.

1.3 Cloud Computing in Remote Sensing

The explosive growth of remote sensing big data has even been revolutionizing the way remote sensing data are managed and analyzed. The significant data volume is far beyond the storage capability of traditional data centers could satisfy, since it is remarkably complicated and expensive to be frequently upgraded and scaled-out to guarantee a growth rate of petabytes per year. Great challenges have also been introduced in the management of the multi-sensor, multi-spectral, multi-resolution and multi-temporal featured large datasets that might be in various formats or distributed across data centers. Likewise, exceptional computational challenges have been routinely introduced by the applications that extend their models to continental or global scale, and even use temporal data for long time-serial analysis. The case could become even worse especially with the spurt of interest in on-the-fly processing that demands tremendous data processing in an extremely short time. Even though the high-performance computing(HPC) resources could mainly facilitate large-scale remote sensing data processing, they are almost out of the reach of many researchers, due to the burden of considerable technical expertise and effort. Accordingly, most existing models hardly accommodate large datasets at global scale, as they are commonly applied on a limited number of datasets within some small area.

Why Cloud computing in remote sensing? Taking advantage of the elasticity and high-level of transparency, Cloud computing offers a revolutionary paradigm that resources are accommodated as ubiquitous services on-demand on a pay-as-use basis along with both flexibility and scalability. Incorporation of Cloud computing to facilitate large-scale scientific applications turns out to be a widespread yet easy-of-use approach to the tremendous computational and storage challenges. In light of this, our book places great emphasis on Cloud-enabled integrated platforms and techniques to empower petabyte-scale remote sensing big data managing, planetary-scale on-the-fly analyzing, together with easy-to-use remote sensing data and processing services on-demand.

Cloud computing in remote sensing, based on virtualization technology, integrates computing, storage, network and other physical resources to build a virtualized resource pool, develops and deploys remote sensing data processing and product production business systems, geographic information comprehensive analysis systems, etc. It provides users with cloud services that integrate data, processing, production, computing platform, storage, and integrated spatial analysis, so as to provide industry application solutions for ecological environment monitoring, land and resources surveys, smart cities, etc.

This book begins with a cloud-based petabyte-scale remote sensing data integration and management infrastructure. Based on an OpenStack-enabled Cloud infrastructure with flexible resources provisioning, a distributed remote sensing data integration system is built to easily integrate petabyte-scale multi-sensor, multi-resolution remote sensing data in various formats or metadata standards across data centers. By virtue of the LSI global data organizing and spatial indexing model as well as distributed No-SQL database, an optimized OODT-based remote sensing data management system is put forward to serve remote sensing data on-demand.

Then, the rest of this book is followed by the introduction of high-performance remote sensing clouds for data centers, namely pipsCloud. It incorporates the Cloud computing paradigm with cluster-based HPC systems for a QoS-optimized Cloud so as to address the challenges from a system architecture point of view, wherein, bare-metal (BM) provisioning of an HPC cluster is proposed for less performance penalty; a data access pattern aware data layout strategy is employed for better data locality and finally optimized parallel data I/O; dynamic DAG scheduling is used for large-scale workflows. Likewise, it benefits from an optimal distributed workflow and resource scheduling strategy on a basis of minimal data transferring, Cloud-enabled remote sensing data producing infrastructure and services are delivered for planetary-scale remote sensing data analyzing across data centers. In addition, the application-oriented generic parallel programming technique is also discussed for easy cloud-empowered large-scale remote sensing data processing. Last but not the least, the remote sensing information analysis, knowledge discovering, remote sensing data and processing service models are all discussed here.

Chapter 2

Remote Sensing Data Integration in a Cloud Computing Environment

2.1 Introduction

Since Landsat-1 first started to deliver volumes of pixels in 1972, the amount of archived remote sensing data stored by data centers has increased continuously [19, 20]. According to incomplete statistics, the total amount of data archived by the Earth Observing System Data and Information System (EOSDIS) reached 12.1 petabytes (PBs) around the year 2015 [21]. Up until August 2017, the archived data volume of the China National Satellite Meteorological Center (NSMC) reached 4.126 PBs [22], and the China Center for Resources Satellite Data and Application (CCRSDA) archived more than 16 million scenes of remote sensing images [23, 24]. Such large amounts of remote sensing data have brought great difficulties for data integration of each data center.

Due to various satellite orbit parameters and the specifications of different sensors, the storage formats, projections, spatial resolutions, and revisit periods of the archived data are vastly different, and these differences have resulted in great difficulties for data integration. In addition, the remote sensing data received by each data center arrives continuously at an ever-faster code rate. It is preferable to ingest and archive the newly received data in order to provide users with the latest data retrieval and distribution service [25]. Therefore, a unified metadata format and a well designed data integration framework are urgently needed.

Hence, for data integration across a distributed data center spatial infrastructure, we proposed an International Standardization Organization (ISO) 19115-based metadata transform method, and then adopted the internationally popular data system framework object-oriented data technology (OODT) [26] to complete the distributed remote sensing data integration.

The rest is organized as follows: Section 2.2 provides an overview of the background knowledge and related work; Section 2.3 describes the distributed multi-source remote sensing metadata transformation and integration; Section 2.4 introduces the experiments and provides an analysis of the proposed program; and Section 2.5 provides a summary and conclusions.

2.2 Background on Architectures for Remote Sensing Data Integration

This section briefly reviews the distributed integration of remote sensing data, as well as the internationally popular data system framework OODT.

2.2.1 Distributed integration of remote sensing data

The most widely used data integration models include [27]:

(1) The data warehouse (DW)-based integration model, which copies all data sources of each heterogeneous database system into a new and public database system, so as to provide users with a unified data access interface. However, due to the heterogeneity of each independent database system, vast data redundancy is generated, and a larger storage space is also required.

(2) The federated database system (FDBS)-based integration model, which maintains the autonomy of each database system and establishes an association between each independent database system to form a database federation, then providing data retrieval services to users. However, this pattern cannot solve the problems of database heterogeneity or system scalability [28].

(3) The middleware-based integration model, which establishes middleware between the data layer and the application layer, providing a unified data access interface for the upper layer users and realizing the centralized management for the lower layer database system. The middleware not only shields the heterogeneity of each database system, providing a unified data access mechanism, but also effectively improves the query concurrency, reducing the response time. Therefore, in this chapter, we will adopt the middleware-based integration mode to realize the distributed remote sensing data integration.

2.2.2 OODT: a data integration framework

An FS or DBMS alone are not suited for the storage and management of remote sensing data. In a "DBMS-FS mixed management mode", remote sensing images are stored in the file system and their metadata are stored and managed by the DBMS. Typical examples are the European Space Agency (ESA) [29], Tiandi Maps of China, the CCRSDA, the NSMC, the China National Ocean Satellite Application Center (NSOAS), and so on. The mixed management mode both effectively solves the quick retrieval and metadata management problems and maintains the high read/write efficiency of the file system. This has been a longtime issue addressed by NASA, whose the Office for Space Science decided to fund the OODT project in 1998.

Apache OODT [30] is an open-source data system framework that is managed by the Apache Software Foundation. OODT focuses on two canonical use cases: big data processing [31] and information integration [32]. It provides three core services: (1) a file manager is responsible for tracking file locations and transferring files from a staging area to controlled access storage, and for transferring their metadata to Lucene or Solr; (2) a workflow manager captures the control flow and data flow for complex processes, and allows for reproducibility and the construction of scientific pipelines; and (3) a resource manager handles allocation of workflow tasks and other jobs to underlying resources, based on the resource monitoring information from Ganglia or other monitoring software.

In addition to the three core services, OODT provides three client-oriented frameworks that build on these services: (1) a file crawler automatically extracts metadata and uses Apache Tika or other self-defined toolkits to identify file types and ingest the associated information into the file manager; (2) a push-pull framework acquires remote files and makes them available to the system; (3) a scientific algorithm wrapper (called the Catalog and Archive Service Production Generation Executive, CAS-PGE) encapsulates scientific codes and allows for their execution, regardless of the environment, while capturing provenance, making the algorithms easily integrated into a production system (Figure 2.1).

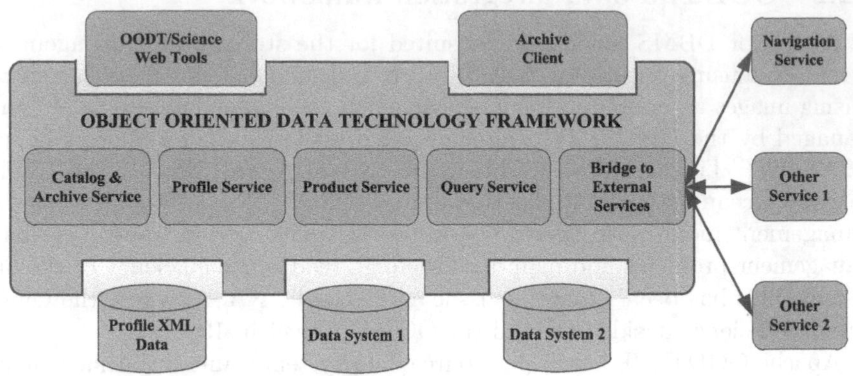

FIGURE 2.1: An object-oriented data technology (OODT) framework.

2.3 Distributed Integration of Multi-Source Remote Sensing Data

With distributed multi-source remote sensing data integration, i.e., based on a unified standard, the remote sensing metadata in the distributed center will be gathered into the main center continuously or at regular intervals, either actively or passively. In this study, the unified satellite metadata standard refers to the ISO 19115-2:2009-based geographic information metadata standard. All of the remote sensing metadata in the distributed sub-centers should be transformed into the ISO 19115-based metadata format before integration to enable uniform data retrieval and management. The distributed sub-centers are mainly responsible for the storage of remote sensing images, and provide an open access interface for the main center based on the HTTP/FTP protocols. The main center is primarily responsible for the ingestion and archiving of the metadata and thumbnails of remote sensing images, and enables uniform query and access for the integrated remote sensing data.

2.3.1 The ISO 19115-based metadata transformation

Remote sensing metadata represent descriptive information about remote sensing images, as well as data identification, imaging time, imaging location, product level, quality, the spatial reference system, and other characteristic information. At present, the metadata forms of different remote sensing data vary greatly. For example, Landsat 8 collects images of the Earth with a

16-day repeat cycle, referenced to the Worldwide Reference System-2 [33]. The spatial resolution of the Operational Land Imager (OLI) sensor onboard the Landsat 8 satellite is about 30 m; its collected images are stored in GeoTIFF format, with Hierarchical Data Format Earth Observation System (HDF-EOS) metadata [34, 35]. The Moderate-Resolution Imaging Spectroradiometer (MODIS) instruments capture data in 36 spectral bands ranging in wavelength from 0.4 μm to 14.4 μm and at varying spatial resolutions (2 bands at 250 m, 5 bands at 500 m, and 29 bands at 1 km). Most of the MODIS data are available in the HDF-EOS format, and it is updated every 1 to 2 days [36]. The charge-coupled device (CCD) sensor, which is carried by the Huan Jing (HJ)-1 mini satellite constellation, has an image swath of about 360 km, with blue, green, red, and near infrared (NIR) bands, 30-m ground pixel resolution, and a 4-day revisit period. Its collected images are stored in GeoTIFF format, and their customized metadata are in eXtensible Markup Language (XML) format [37]. These different metadata formats have resulted in great difficulties for data integration and management, which could be solved by transforming them into a uniform metadata format for uniform retrieval and management [38, 39].

ISO 19115-2:2009 is the geographic information metadata standard which was published by the International Standardization Organization (ISO). It mainly defines the metadata schema of geographic information and services, including the identification, quality, space range, time horizon, content, spatial reference system, distribution, and other characteristic information [40]. Currently, ISO 19115-2:2009 has been integrated into the Common Metadata Repository (CMR) as one of the most popular standards for data exchange [41], data integration, and data retrieval across international geographic information organizations and geographic data centers.

On the basis of the ISO 19115-2:2009 geographic information standard, we proposed a uniform remote sensing metadata format. All of the remote sensing metadata in the distributed sub-centers should be transformed into this uniform format before data integration. In this chapter, the transformational rules we established are mainly aimed at NASA EOS HDF-EOS format metadata (Aster and Landsat series satellites included) and the customized XML-based metadata of the CCRSDA (HJ-1A/B, GF and ZY series satellites included) (see Table 2.1).

It should be noted that in Table 2.1, the strike-through (-) shows the field does not exist, and it will be assigned a null value after metadata transformation. In the ISO metadata column, the term spatialResolution describes the ability of the remote sensor to distinguish small details of an object, generally in meters, thereby making it a major determinant of image resolution. Hence, the spatialResolution is mapped to NadirDataResolution in the HDF-EOS metadata column and pixelSpacing in the CCRSDA metadata column. The terms scenePath and sceneRow are orbit parameters of the satellite in the Worldwide Reference System (WRS), just mapping to WRS_PATH and WRS_ROW in the HDF-EOS metadata column. The term imageQualityCode is a characteristic of a remote sensing image that measures the perceived image degradation, and has the same meaning as the overallQuality in the CCRSDA metadata column. The term processingLevel denotes the type of the remote sensing data, and is mapped to the DATA_TYPE in the HDF-EOS metadata column and productLevel in the CCRSDA metadata column.

TABLE 2.1: The ISO 19115-2:2009-based uniform metadata format and transformational rules. ISO: International Standardization Organization; CCRSDA: China Center for Resources Satellite Data and Application; HDF-EOS: Hierarchical Data Format Earth Observation System.

Categories	ISO Metadata	HDF-EOS Metadata	CCRSDA Metadata
Metadata information	Creation	FILE_DATE	-
	LastRevision	-	-
Image Information	MD_Identifier	LOCALGRANULEID	-
	TimePeriod_beginposition	RangeBeginningDate+RangeBeginningTime	imagingStartTime
	TimePeriod_endPosition	RangeEndingDate+RangeEndingTime	imagingStopTime
	Platform	AssociatedPlatformShortName	satelliteId
	Instrument	AssociatedinstrumentShortName	-
	Sensor	AssociatedsensorShortName	sensorId
	Datacenter	PROCESSINGCENTER	-
	recStationId	STATION_ID	recStationId
	spatialResolution	NADIRDATARESOLUTION	pixelSpacing
	westBoundLongitude	WESTBOUNDINGCOORDINATE	productUpperLeftLong
	eastBoundLongitude	EASTBOUNDINGCOORDINATE	productUpperRightLong
	southBoundLatitude	SOUTHBOUNDINGCOORDINATE	productLowerLeftLat
	northBoundLatitude	NORTHBOUNDINGCOORDINATE	productUpperLeftLat
	centerLongtiude	-	sceneCenterLong
	centerLatitude	-	sceneCenterLat
	scenePath	WRS_PATH	scenePath
	sceneRow	WRS_ROW	sceneRow
	referenceSystemIdentifier	PROJECTION_PARAMETERS	earthModel+mapProjection
	cloudCoverPercentage	-	cloudCoverPercentage
	imageQualityCode	-	overallQuality
	processingLevel	DATA_TYPE	productLevel

2.3.2 Distributed multi-source remote sensing data integration

Distributed multi-source remote sensing data integration refers to the process of validating, inserting, updating, or deleting metadata in the main center metadata management system; it affects only the metadata for the distributed data providing sub-centers. The metadata management is mainly realized by the components of OODT, including the OODT crawler, OODT push-pull, and OODT file manager [42] (see Figure 2.2).

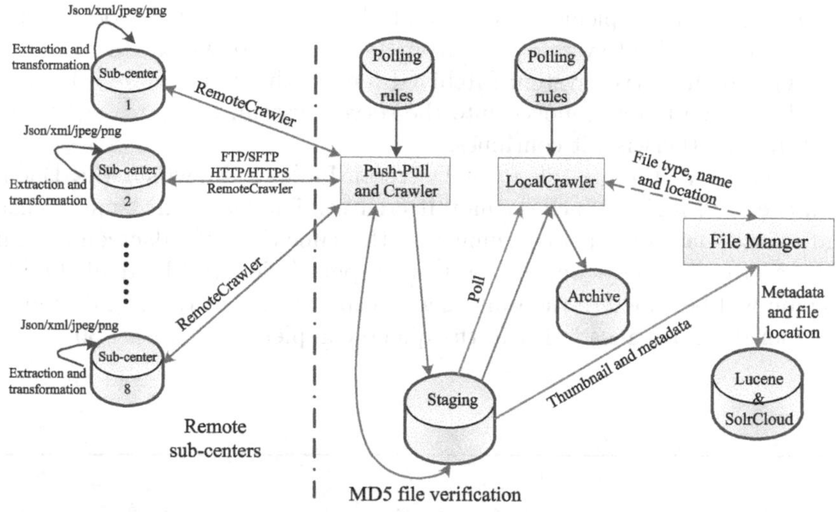

FIGURE 2.2: The process of distributed data integration.

In the main data center, the push-pull daemon will be launched automatically by using its daemon launcher at the defined time interval. The daemon will wrap one of two processes: (1) RemoteCrawler, or (2) ListRetriever. The RemoteCrawler process crawls remote sites for files in the distributed sub-centers. Meanwhile, the RemoteCrawler process also automatically extracts metadata and transforms them into the ISO 19115-2:2009-based uniform metadata format. The ListRetriever retrieves known files from remote sites in the distributed sub-centers (that is, the path and file name to each file is known and has been specified in a property file, and a parser for that property file has been specified). After crawling or retrieval, the push-pull framework will be responsible for downloading remote content (pull), or accepting the delivery of remote content (push) to the main center for use by the LocalCrawler for ingestion into the file manager. Here, the remote content includes the metadata file and thumbnail of remote sensing data. It is worth mentioning that the LocalCrawler is developed in the main center, and is primarily responsible for crawling the local client system for files in the main center. The file manager component is responsible for tracking, ingesting, and moving metadata and thumbnails between a client system and a server system in the main center. Finally, the remote sensing metadata will be indexed by the SolrCloud, and their corresponding thumbnails will be archived in the file system.

Both the RemoteCrawler and LocalCrawler have an incremental control mechanism in order to avoid duplicate data ingestion. In the intervals between crawling and data ingestion, the RemoteCrawler executes a Message Digest 5 (MD5) file verification process between the remote sites' files in the sub-center and the archived files in the main center. If the file has been archived in the main center, data ingestion will be stopped; otherwise, data ingestion continues. The LocalCrawler implements the second MD5 file verification process between the files in the client system (files from sub-centers downloaded to the main center) and the server system (archived files in the main center). If the files have been ingested and moved into the server system, the data ingestion will be stopped; otherwise, it continues.

In addition, there is also the DaemonManager, in which the DaemonLauncher will register each daemon it creates. The DaemonManager ensures that no two Daemons are ever running at the same time. If a daemon is running when another requests permission to run, permission will be denied and the daemon will be added to the wait queue until the current running daemon and all other daemons ahead of it in the queue complete their tasks [30].

2.4 Experiment and Analysis

In order to verify the availability of our proposed solution, a virtual multi-data center environment was set up based on the OpenStack cloud computing framework. The main data center was composed of three Linux virtual machines. All of the three machines were developed with the SolrCloud environment, responsible for metadata index and retrieval. One of them was developed with OODT system framework, responsible for data ingestion and thumbnail archiving. The distributed sub-center was composed of eight Linux virtual machines, corresponding to eight satellite data centers. Each machine was mounted with a one-terabyte (TB) cloud drive so as to provide image storage space. In addition, all the machines in the main and sub centers were configured with 4 gigabytes (GBs) of RAM and 2 virtual processor cores. The framework of the virtual multi-data center environment is shown in Figure 2.3.

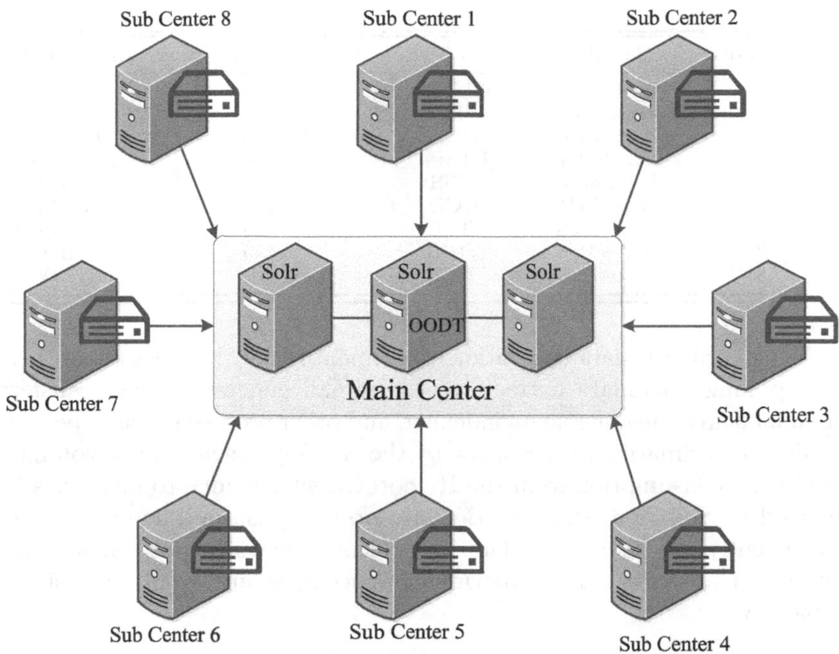

FIGURE 2.3: The framework of the virtual multi-center data environment.

The experimental images of the distributed integration test mainly include Landsat 8 OLI_TIRS, Landsat 7 ETM+, Landsat 5 TM, Landsat 1–5 MSS, Aster L1T, CEBERS-1/2 CCD, HJ-1A/B CCD, HJ-1A HSI, and FY-3A/B VIRR images, which were freely downloaded from the USGS (`https://earthexplorer.usgs.gov/`), NSMC (`http://satellite.nsmc.org.cn/portalsite/default.aspx`) and CCRSDA (`http://www.cresda.com/CN`) websites. A total of 3380 files were downloaded. These images were distributed in the eight sub-centers according to data type. The total number of our experimental images are shown in Table 2.2.

TABLE 2.2: A summary of the experimental images.

Sub-Center	Satellite	Data Type	Volume of Images	Image Format
1	Landsat 8	OLI_TIRS	310	GeoTIFF
2	HJ-1A	HSI	350	HDF5
2	CEBERS-1/2	CCD	270	GeoTIFF
3	Landsat 7	ETM+	450	GeoTIFF
4	Landsat1-5	MSS	260	GeoTIFF
5	HJ-1A/B	CCD	710	GeoTIFF
6	Landsat 5	TM	430	GeoTIFF
7	FY-3A/B	VIRR	450	HDF5
8	Aster	L1T	150	HDF4

The distributed data integration experiment mainly includes remote sensing data polling, metadata extraction, thumbnail generation, file transferring, thumbnail archiving, metadata indexing, and other processes. The experimental results are primarily with respect to the already-crawled data volume and total time consumption from the RemoteCrawler launch to metadata being indexed by SolrCloud/Lucene. Because no two push-pull daemons ever run concurrently, the distributed data integration experiment was carried out one sub-center at a time. The experiment procedures and results are shown in Table 2.3.

TABLE 2.3: Experimental results of distributed data integration.

Satellite	Data Type	Volume of Images Stored in Sub-Center	Volume of Images Integrated by Main Center	Average Transfer Rate(MB/s)
Landsat 8	OLI_TIRS	310	310	9.8
HJ-1A	HSI	350	350	10.1
CEBERS-1/2	CCD	270	270	11.7
Landsat 7	ETM+	450	450	10.5
Landsat1-5	MSS	260	260	12.8
HJ-1A/B	CCD	710	710	9.9
Landsat 5	TM	430	430	13.8
FY-3A/B	VIRR	450	450	11.2
Aster	L1T	150	150	10.8

As can be seen in Table 2.3, the number of main center-integrated remote sensing images is equal to the total number of each sub-center's stored images. That is to say, there is no information lost during the process of data integration. Moreover, our designed ISO 19115-2:2009-based uniform metadata model includes all fields of integration by participating remote sensing metadata, and the SolrCloud indexed metadata can also maintain the metadata information of each remote sensing image perfectly. As for the transfer rate, it mainly depends on the window size for the OODT-push-pull component. In our experiment, the window size was set at 1024 bytes, and the average transfer rate is between 9.8 and 13.8 MB/s. This is enough to satisfy the demands of metadata and thumbnail transfer across a distributed data center spatial infrastructure.

Therefore, the experimental results showed that our OODT-based distributed remote sensing data integration was feasible.

2.5 Conclusions

In view of the current issues of remote sensing data integration, we proposed an OODT-based data integration framework. Specifically, aiming at heterogeneous features of multi-source remote sensing data, we proposed an ISO 19115-2:2009-based metadata transform method to achieve unity of metadata format in the distributed sub-centers. In order to achieve efficient, stable, secure and usable remote sensing data integration across a distributed data center spatial infrastructure, we adopted the OODT framework based on its stable, efficient, and easy-to-expand features, to implement remote sensing data polling, thumbnail generation, file transfer, thumbnail archiving, metadata storage, etc. In addition, in order to verify the availability of our proposed program, a series of distributed data integration experiments was carried out. The results showed that our proposed distributed data integration program was effective and provided superior capabilities.

However, the unified metadata conversion rule was pre-configured, and the metadata transformation was done manually. This was convenient and easy to operate, but less efficient. In particular, with an increase of data types, a great burden would be brought to data integration. Future studies based on deep learning algorithms using semantic matching and unified format conversion of remote sensing metadata will be performed.

Chapter 3

Remote Sensing Data Organization and Management in a Cloud Computing Environment

3.1 Introduction

With the unprecedented development in sensor technologies, the remote sensing data are recognized as big data due to its significant growth in volume, velocity and variety. For example, the Earth Observing System Data and

Information System (EOSDIS) developed by the National Aeronautics and Space Administration (NASA) holds 14.6 PBs archived data as of September 30, 2015, and the average daily archive growth is 16 TB per day [43]. In China, by February 2018, the total volume of archived remote sensing data in the China National Satellite Meteorological Centre (NSMC) has reached 4.59 PBs [44]. Additionally, the improved data resolution in time, space and spectrum also leads to tremendous growth in remote sensing data size and volume [45]. On the other hand, the remote sensing data are often delivered from various platforms carrying at least one sensor to different ground stations that are geographically distributed [46]. Hence, the remote sensing data have the characteristics of multi-sources, multi-temporal and multi-resolution [47], resulting in the variety of remote sensing data in terms of its data types and data formats.

Generally, it is preferable to gather distributed remote sensing data and make them available for the public [48]. However, the proliferation of remote sensing data poses new challenges for efficient remote sensing data management and retrieval. But retrieval efficiency is mainly dependent upon the data organization model and storage system.

Currently, the two most widely used data organization models are: (1) spatio-temporal recording system-based satellite orbit stripes or scene organization; and (2) globally meshed grid-based data tiling organization [49]. However, the former has obvious shortcomings for massive data retrieval and quick access; and the latter causes an increase by about one-third in the amount of data due to image segmentation, thus requiring larger data storage spaces. Hence, we should explore a reasonable spatial organization mode for massive, multi-source remote sensing data, so as to improve the efficiency of massive data retrieval.

As for the data storage system, the combination of the file system (FS) and the database management system (DBMS) is the mostly popular mode in the management of remote sensing data and its metadata [48]. The "DBMS-FS mixed management mode" solves both the problems of metadata management and quick retrieval, and also maintains the high read/write efficiency of a file system. In the context of the rapid growing flood of remote sensing data, distributed storage systems offer a remarkable solution to meet the challenges caused by the significant growth of remote sensing data volume and velocity. Therefore, how to organize the massive remote sensing data and its metadata in distributed storage systems based on the "FS-DBMS" management mode is another problem that needs to be solved. Therefore, we proposed a logical segmentation indexing (LSI) model to achieve the organization of integrated remote sensing metadata, and chose SolrCloud to realize the distributed index and quick retrieval [50]. The LSI model takes the logical segmentation indexing code as the identifier of each remote sensing data, rather than performing an actual physical subdivision. This not only increases the efficiency of data retrieval with the help of the global subdivision index, but also avoids generating numerous small files caused by the physical subdivision of data.

Futhermore, recently, Hadoop and its ecological environment have become the most popular platform used for big data storage and processing [16]. Lots of landmark works have been conducted to enable remote sensing big data processing in Hadoop platform [51, 52, 53, 54]. However, Hadoop is primarily designed for processing a large scale of web data; it does not natively support commonly used remote sensing data formats such as GeoTiff, HDF, NetCDF. There are two approaches to enable remote sensing big data processing in Hadoop platform. One feasible approach is to convert the remote sensing data to Hadoop friendly data formats. For example, both [55] and [56] divide the remote sensing data into a number of blocks, each of which is stored in HBase and is processed by the MapReduce paradigm. The other approach is to develop additional plugins to enable Hadoop supporting the commonly used remote sensing data formats. For example, In [54], researchers modify the `RecordReader` interface to enable MapReduce processing Geos and GeoTiff formats remote sensing data.

However, current works mostly concentrate on the management or the processing of remote sensing data in Hadoop and its ecological environment, and do not take the heterogeneity of remote sensing data into consideration. Therefore, in this study, a scalable, efficient, distributed data management system is proposed to store and index heterogeneous remote sensing data. The system is composed of two main components. The Data Organization and Storage Component is responsible for organizing and storing the remote sensing data and its metadata in Hadoop and its ecological environment. The Data Index and Search Component creates the spatial index and full-text index based on remote sensing metadata contents, and provides different search types for users. Besides, two MapReduce-based bulkload techniques are adopted to improve the speed of data loading.

The remainder of this chapter is organized as follows. Section 3.2 introduces preliminaries and related techniques. The detailed implementations of the proposed LSI organization model are presented in Section 3.3. Section 3.4 and 3.5 describe the data management mode in parallel file system and Hadoop ecosystem respectively, and their corresponding experiments are detailed in Section 3.6 and 3.7. Finally, Section 3.8 concludes this chapter.

3.2 Preliminaries and Related Techniques

3.2.1 Spatial organization of remote sensing data

The main two models for spatial organization of multi-source remote sensing data are: (1) the satellite orbit stripe or scene organization based on the spatio-temporal recording system; and (2) data tiling organization based on the globally meshed grid [57, 58].

In the first model, the original orbit data are organized according to reception time, and they are stored in a scene unit. Each scene unit is identified by upper, lower, left and right four-point latitude and longitude coordinates. This simple organization method has been adopted by remote sensing data centers around the world, such as NASA's Earth Observing System (EOS) [59] and the CCRSDA [49]. However, due to the four-point index queries in database systems, this model has obvious shortcomings for massive data retrieval and quick access. Therefore, in this chapter, we have proposed the LSI model to reduce the dimension of the query index, and this will be described in Section 3.3.

In the second model, remote sensing images are subdivided into buckets of grid shape, and each bucket is labeled by a unique geocode according to certain coding rules. This is especially useful in database systems where queries on a single index are much easier or faster than multiple-index queries. Furthermore, this index structure can be used for a quick-and-dirty proximity search: the closed points are often among the closest geocodes. The longer a shared prefix is, the closer the two buckets are [60]. This model is generally applicable to the image cache systems and map publishing systems typically used by Google Earth, Bing Maps, and Tiandi Maps of China, for example [61]. However, due to image segmentation and pyramid construction, this model means the amount of data increases by approximately one-third, so that a larger data storage space is required; it also generates a large number of small tiles, which can easily cause a single point of failure, and are not conducive to data storage and management using the distributed file system [62]. Hence, in this chapter, we proposed a logical partition index and virtual mapping construction strategy for scene-based remote sensing data, and this will be also described in Section 3.3.

3.2.2 MapReduce and Hadoop

MapReduce [12] is proposed by Google and has emerged as the de facto paradigm for large scale data processing across a number of commodity machines. The procedures in MapReduce can be divided into the Map stage and the Reduce stage. In the Map stage, the data is divided into a number of data splits; each of them is processed by a `mapper` task. The `mapper` task reads data from its corresponding data split and generates a set of intermediate key/value pairs. These intermediate key/value pairs are then sent to different partitions; each partition is processed by a `reducer` task in the Reduce stage, and generates the final results. To enhance the performance of MapReduce applications, an optional procedure called `Combiner` is often used to gather key/value pairs with the same key. The number of intermediate key/value pairs processed by `reducer` tasks can be significantly alleviated after the `Combiner` is executed, resulting in efficiency improvement.

Hadoop [13] is an open-source implementation of Google's MapReduce paradigm. As one of the top-level Apache projects, Hadoop has formed a Hadoop-based ecosystem [15], which becomes the cornerstone technology of many big data and cloud applications [16]. Hadoop is composed of two main components: the Hadoop Distributed File System (HDFS) and the MapReduce paradigm. HDFS is the persistent underlying distributed file system, and is capable of storing terabytes and petabytes of data. The data stored in HDFS are divided into fixed-size (e.g. 128 MB) data chunks, each of which has multiple replicas (normally three replicas) across a number of commodity machines. The data stored in HDFS can be processed by the MapReduce paradigm.

3.2.3 HBase

HBase [63] is a scalable, reliable, distributed, column-oriented big database. As a component of the Hadoop-based ecosystem, HBase is tightly coupled with Hadoop since it normally adopts HDFS as its underlying file system. Besides, HBase also provides the MapReduce programming interfaces to read/write data stored in HBase.

The data stored in HBase are organized as tables. Each table contains a large number of sorted rows, each of which has a unique row key and a number of column families. There are different numbers of qualifiers within each column family, and at the intersections of rows and column qualifiers are the table cells. Table cells are used to store actual data and often have multiple versions specified by timestamp. A table cell can be unique identified by the following sequence: (Table, Row Key, Column Family: Column Qualify, Timestamp). Tables stored in HBase are horizontally split into a number of regions, each of which is assigned to specific regions determined by HBase Master. Regions assigned to each region server are further vertically divided into many files stored in HDFS according to the column families.

HBase is the emerging NoSQL database based on Hadoop, and provides random real-time read/write access to big data. However, HBase only supports two basic lookup operations based on the row key: exact match and range scan. There is no other native mechanism for cells data searching [64]. To solve this problem, this chapter builds a secondary index based on Elasticsearch to provide additional searching abilities for cell contents.

3.2.4 Elasticsearch

Elasticsearch is an open-source, distributed, RESTful, real-time full-text search engine based on Apache Lucene [65]. Elasticsearch is designed to store, index and search a large sets of documents. Documents are converted to JSON objects and are stored in Elasticsearch indices. Each Elasticsearch index is composed of one or more Lucene indices called primary shards. Normally, in a Elasticsearch cluster, each primary shard often has zero or more replica shards. By this means, Elasticsearch can automatically balance the loads between available shards in the cluster to improve the overall performance.

Figure 3.1 depicts an Elasticsearch cluster. This Elasticsearch cluster has one master node and two data nodes. There is an index called `rsIndex` with four primary shards in the cluster. Each primary shard has only one replica shard. When users commit a document added request to the Master node, it

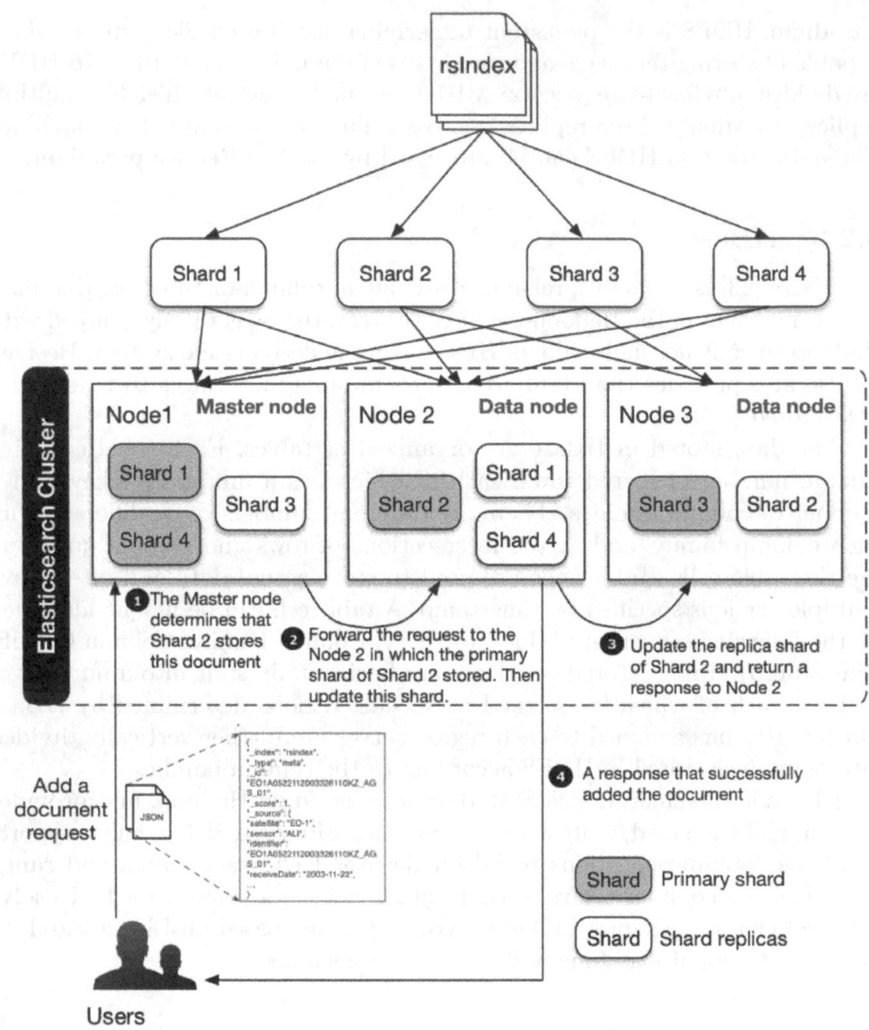

FIGURE 3.1: Example of an Elasticsearch cluster.

first calculates which shard the new document belongs to. Once the shard is determined, the request is forwarded to the node storing the primary shard. In the example, Node 2 stores the primary shard of Shard 2 and handles the request to add the new document into the primary shard. Then, the request is sent to nodes storing the replica shards of Shard 2. In this example, only Node 3 holding the replica of Shard 2 needs to handle the request. Once the document is added into all shards of Shard 2 successfully, users will receive a response sent by Node 2 to identify that the new document is added into the cluster successfully.

Elasticsearch has shown significant performance in storing, indexing and searching a large scale of documents in the big data environment. Recently, Elasticsearch has been natively integrated with Hadoop and its ecosystem [66]. A plugin called `Elasticsearch-Hadoop`[1] is specifically developed to let Elasticsearch read/write data from/to HDFS. However, there is no native support for the integration between Elasticsearch and HBase. Therefore, this chapter develops the interfaces between Elasticsearch and HBase. The data stored in HBase are loaded into the Elasticsearch cluster with the MapReduce paradigm. Besides, to improve the speed of data loading, a MapReduce-based bulkload technique is adopted.

3.3 LSI Organization Model of Multi-Source Remote Sensing Data

The LSI organization model is based on the Geographical Coordinate Subdividing Grid with One Dimension Integer Coding on 2^nTree (GeoSOT) grid, which was proposed by the research group of Cheng around 2012 [67]. The main idea is expansion by three times for the latitude and longitude of Earth's surface. The first expansion is the original $180° \times 360°$ Earth surface extended to $512° \times 512°$; the expanded surface is viewed in level 0 grids, with grid code 0. Then, the level 0 grid is recursively partitioned quadrilaterally until reaching the 1° grid cell, with a total of nine subdivisions. The second expansion is processed for the 1° grid cell, namely, 1° extended to $64'$. The extended $64'$ grid cell is recursively partitioned quadrilaterally until reaching the $1'$ grid cell, with a total of 12 subdivisions. Similarly, the $1'$ grid cell is recursively partitioned quadrilaterally until reaching the $\frac{1''}{2048}$ grid cell, with a total of 11 subdivisions.

Finally, after the three expansions and 32 subdivisions, the system is used to cover the whole world, dividing the Earth into centimeter-level units using a hierarchy grid system with whole degrees, whole minutes, and whole seconds. Taking 32-bit quaternary coding at the Z-sequence, level subdivision cells were named as 00, 01, 02, 03, and so on; the location relationship of various spatial information products in different coordinate systems can be built with these globally unique and geographically meaningful codes [68, 69] (Figure 3.2).

Based on the GeoSOT global segmentation strategy, the logical partition indexing code of each scene-based remote sensing data was calculated first in this chapter. It is worth noting that there are three cases to consider regarding the logical code in general. Firstly, when the minimum bounding rectangle (MBR) [70] of a remote sensing image is completely contained in a GeoSOT grid, the logical partition index code is the corresponding GeoSOT grid code. Secondly, when the MBR of a remote sensing image spans two grids, the two

[1]`https://github.com/elastic/elasticsearch-hadoop`

grid codes will be the logical partition index codes. Thirdly, when the MBR of a remote sensing image spans four grids, the logical partition codes will be composed of the four codes [56] (Figure 3.3).

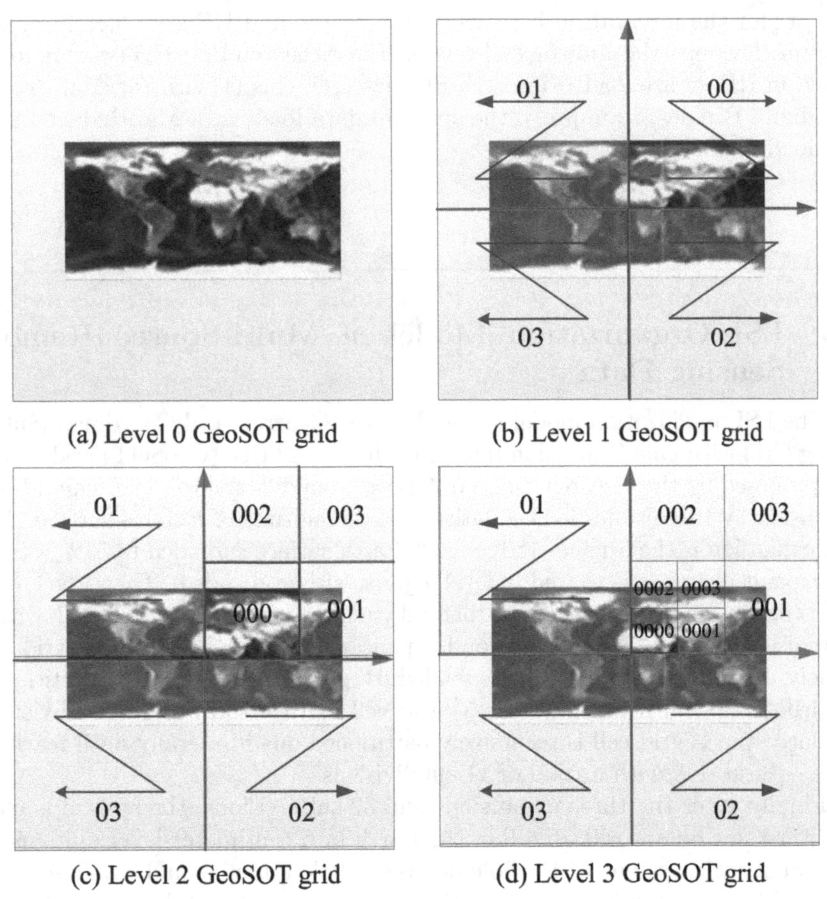

FIGURE 3.2: Level 0-3 Geographical Coordinate Subdividing Grid with One Dimension Integer Coding on 2^nTree (GeoSOT) grids.

After encoding each type of scene-based remote sensing data, then the virtual mapping between the logical partition indexing codes and position parameters (latitude and longitude) of each scene can be established easily (Figure 3.4). In fact, the logical partition indexing codes have become the form of spatial identification of each piece of scene-based remote sensing data when the virtual mapping is created. Reducing 8-index queries (latitude and longitude values of the upper, lower, left, and right four points) to no more than 4-index queries, the query speed increase in database systems is obvious.

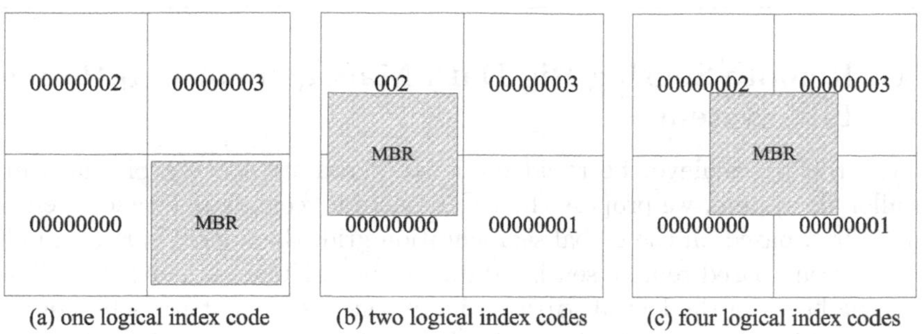

FIGURE 3.3: Three cases with respect to the logical partition index code. MBR: minimum bounding rectangle.

In addition, the logical partition indexing code of each scene center point is always used in the actual query process, and just one index query could be made in this situation. Therefore, based on the center point indexing code, a quick retrieval of massive remote sensing data can be realized.

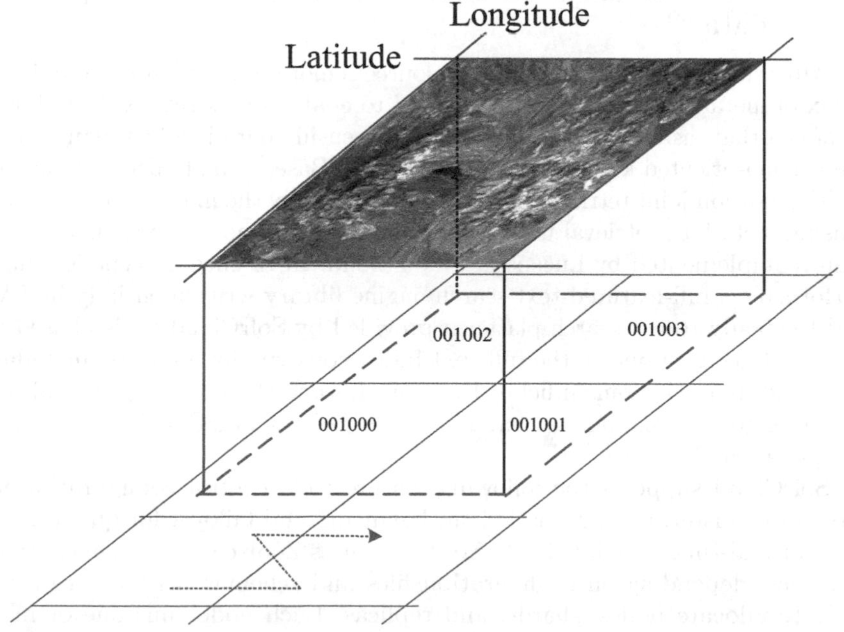

FIGURE 3.4: The virtual mapping between the logical partition indexing and the scene parameters of remote sensing data.

3.4 Remote Sensing Big Data Management in a Parallel File System

In order to achieve the rapid retrieval of remote sensing big data in a parallel file system, we propose the LSI model for scene-based remote sensing data. First, based on the global segmentation grid, the logical partition index of each scene-based remote sensing data can be calculated. Then, the virtual mapping between the logical partition index and the scene parameters of each remote sensing data can also be established easily. Finally, based on the logical partition index and virtual mapping, as well as the full-text search engine Solr/SolrCloud, quick retrieval of remote sensing data becomes possible. The LSI model-based data retrieval not only improves data query efficiency with the help of the global subdivision index encoding, but also avoids generating small files caused by the actual data subdivision. This section will describe the LSI model and SolrCloud-based remote sensing metadata management in terms of the LSI model, full-text index construction, and distributed data retrieval, etc.

3.4.1 Full-text index of multi-source remote sensing metadata

After spatial organization of multi-source remote sensing data, the full-text index of metadata should be constructed to enable quick retrieval. It should be added that, as the query index of remote sensing data involves many terms, the column-oriented key-value data store, like HBase, cannot effectively handle multi-condition joint retrieval. Hence, in this chapter, the multi-sourced remote sensing metadata retrieval used the full-text index, and its construction was mainly implemented by Lucene and SolrCloud. In essence, Lucene is a high-performance, full-featured text search engine library written entirely in Java, and the ready-to-use search platform provided by SolrCloud is also based on Lucene. Lucene supports the full-text index construction of static metadata fields and dynamic domain fields. However, Lucene is not a complete full-text search engine; it should be combined with Solr or SolrCloud to provide a complete search service [71].

SolrCloud supports the following features: (1) central configuration for the entire cluster; (2) automatic load balancing and failover for queries; and (3) near real-time search [72, 73]. SolrCloud uses ZooKeeper to manage these locations, depending on configuration files and schemas, without a master node to allocate nodes, shards, and replicas. Each node runs one or more collections, and a collection holds one or more shards. Each shard can be replicated among the nodes. Queries and updates can be sent to any server. Solr uses the information in the ZooKeeper database to figure out which servers need to handle the request. Once the SolrCloud cluster starts, one of the nodes is selected as a leader, which is responsible for all shards [74]. In addition, there is a master controller in the cluster, called the overseer node, which is responsible for maintaining cluster state information and thereby provides for failover to the Solr cluster (Figure 3.5).

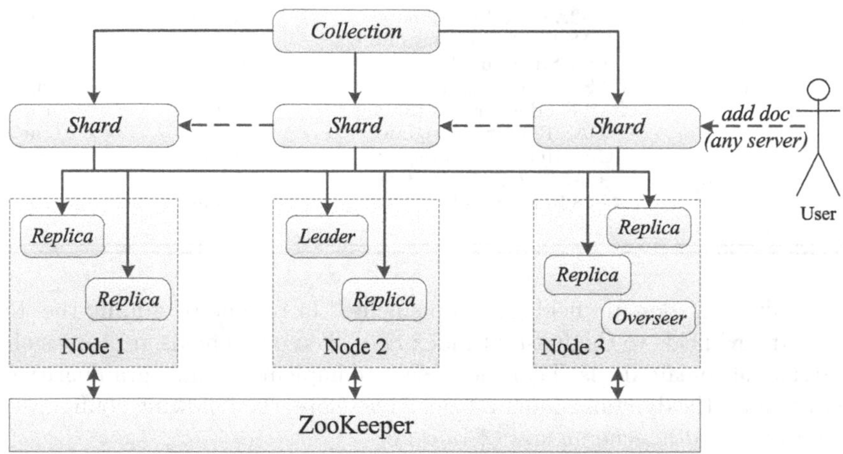

FIGURE 3.5: SolrCloud.

In this study, the distributed index based on SolrCloud/Lucene was applied on the static and dynamic metadata fields. As the name suggests, the distributed index will be used when our index collections are so large that we cannot construct an index efficiently on a single machine. The static remote sensing metadata fields are defined by the OODT file manager, and include ID, CAS.ProductId, CAS.ProductTypeName, CAS.ProductReceivedTime, and CAS.ProductTransferStatus, CAS.ProductName, CAS.ProductStructure, and so on. The dynamic domain fields mainly include the satellite type, sensor type, scanning time, and GeoSOT codes. In addition, in order to conduct the comparative experiments, the original latitude and longitude of each image are also included in the dynamic domain fields. The static and dynamic fields' index is as shown in Table 3.1. It is noted that in Table 3.1, the asterisk (*) denotes all of the dynamic domain fields of the remote sensing metadata.

TABLE 3.1: The full-text index structure of multi-source remote sensing metadata.

IndexType	Field	FieldType	Indexed
static	ID	string	true
	CAS.ProductId	string	true
	CAS.ProductName	string	true
	CAS.ProductTypeName	date	true
	CAS.ProductTypeId	string	true
	CAS.ProductReceivedTime	string	true
	CAS.ProductTransferStatus	string	true
	CAS.ReferenceOriginal	string	true
	CAS.ReferenceDatastore	string	true
	CAS.ReferenceFileSize	long	true
	CAS.ReferenceMimeType	string	true
dynamic	*	string	true

The dynamic domain fields are implemented in Lucene by adding the 'text' and 'text_rev' fields to the full-text index field. These two fields are copies of all the dynamic domain fields. Their purpose is to implement multi-granularity segmentation for the dynamic domain fields. Therefore, the following configuration should be added in schema.xml of Lucene.

copyField source="*" dest="text"
copyField source="*" dest="text_rev"

3.4.2 Distributed data retrieval

After the construction of the distributed full-text index in Lucene, the index will be partitioned across several machines. Hence, data retrieval will be executed on several machines, and realized by the distributed search server SolrCloud. In the SolrCloud distributed clusters, all full-text indexes can make up a collection comprising one logical index. The collection is usually split into one or more shards, and evenly distributed on each node based on routing rules. In general, all shards in the same collection have the same configuration. Each shard usually has one or more replicas; one replica of each shard will be elected as a leader [75, 76]. In this study, the collection was split into three shards, and each shard had three replicas.

In addition, there is an overseer node in the cluster that is responsible for maintaining cluster state information. It will monitor the status of each Leader node, acting as a master controller. When one shard's leader falls offline, the overseer node will initiate the automatic disaster recovery mechanism, and another node in the same shard will be designated as the leader to provide service. Even if the overseer node fails, a new overseer node will be automatically enabled on another node, ensuring high availability of the cluster. In the meantime, the index replica on the off-line node will be automatically rebuilt and put to use on other machines.

The retrieval of distributed metadata in SolrCloud is implemented as follows: once any one of the SolrCloud nodes receives a data query request, the request will be forwarded to one of the replication nodes by the internal processing logic of the cluster. Then the replication node will launch the distributed query according to the created full-text index of remote sensing data. The distributed query will be converted into multiple sub-queries, each of which will be located on any of the replications of their corresponding shard. It is worth noting that the number of sub queries is equal to the number of shards. Finally, the results of each sub-query will be merged by the replication node that received the original query, and the merged final query results will be returned to the user. In addition, automatic load balancing is also provided by SolrCloud. If the query pressure is too large, the cluster scale can be expanded and replications increased to smooth the query pressure. The SolrCloud distributed query process is shown in Figure 3.6.

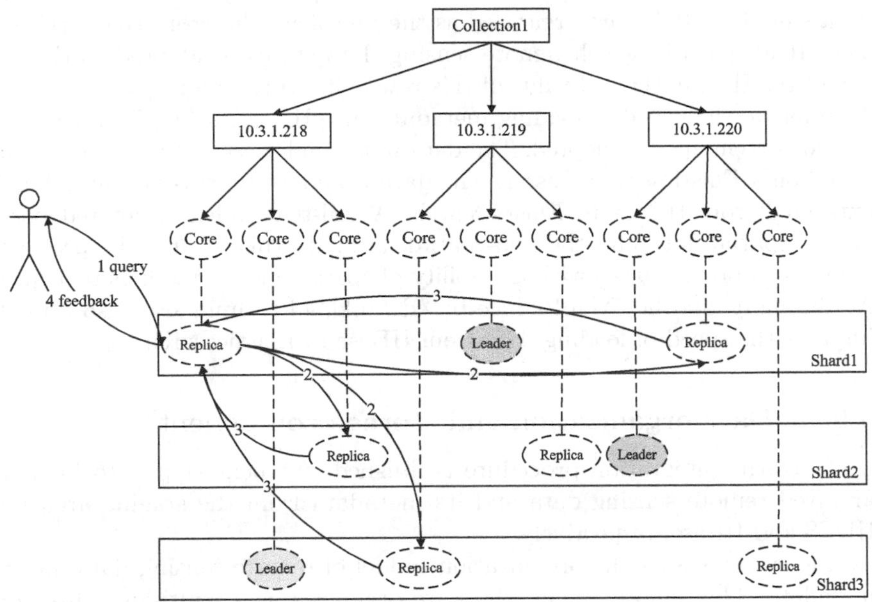

FIGURE 3.6: The SolrCloud distributed query process. (1) The user's data query request is sent to any one of the SolrCloud nodes, and then forwarded to one of the replication nodes; (2) the distributed query is launched and converted into multiple sub-queries, each of which is located on any of the replications; (3) results are returned by each sub-query; and (4) sub-query results are merged and returned to users.

3.5 Remote Sensing Big Data Management in the Hadoop Ecosystem

Figure 3.7 presents an overview of the system. The system is composed of three main components. The data integration component automatically, periodically extracts remote sensing data and its metadata from distributed data sources, transforms the data and metadata to form a unified data view and loads both data and metadata from different data sources into the proposed system. The key to the data integration component is how to solve the heterogeneity of remote sensing data and form a unified view. Once the remote sensing data are gathered from different data sources into the staging area of the proposed system, the remote sensing data and its metadata are loaded into HDFS and HBase for persistent storage, respectively. The remote sensing data stored in HDFS are organized as the multilevel directory tree structure. The HDFS path in which remote sensing data stores is adopted as the row key of the HBase, the cell value of this row is its corresponding metadata file. To improve the speed of loading metadata into HBase, a MapReduce-based bulkload technique is adopted. The data index and search component is built based on a Elasticsearch cluster. The data of indexed terms are periodically extracted from HBase to Elasticsearch. A full-text index is created based on the spatial data and the non-spatial data. By this means, the proposed system can provide users with the ability of spatial query as well as text query. Additionally, another MapReduce-based bulkload technique is also built to improve the speed of loading data from HBase to Elasticsearch.

3.5.1 Data organization and storage component

Once the integration procedure is finished, the next step is to load the archived remote sensing data and its metadata from the staging area into HDFS and HBase, respectively.

Figure 3.8 shows the organization model of remote sensing data and its metadata. The remote sensing data are organized as a multi-level directory tree in HDFS based on the order Satellite/Sensor/Year/Month/Day. In this way, each remote sensing data has a unique storage path in HDFS. The storage path is adopted as the row key of HBase. Meanwhile, The cell value of this row is its corresponding metadata.

Loading a large number of small metadata into HBase is time consuming work. To improve the speed of the metadata loading procedure, a MapReduce-based bulkload technique is adopted in the system. The process of the bulkload technique is illustrated in Figure 3.9. The archived metadata are first loaded into a temporary directory in HDFS. Then a MapReduce program is executed. In the Map stage, a number of `mappers` read the metadata from HDFS and generate a set of HBase operation sets. Then the `reducers` execute the

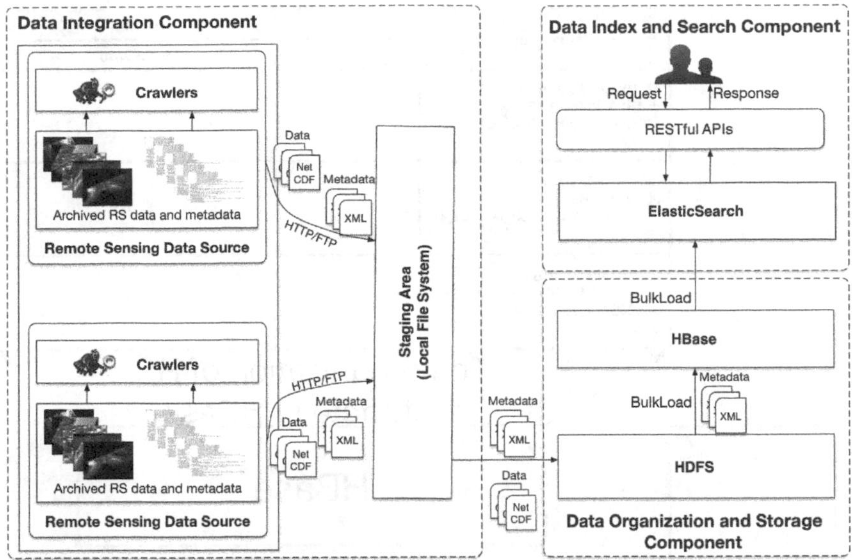

FIGURE 3.7: System architecture for heterogeneous remote sensing data integration and management.

operation sets, generate HFiles and store them in a HDFS directory. Once the MapReduce progress is finished, the The bulkloader is launched to allocate the generated HFiles to different Region Servers.

3.5.2 Data index and search component

HBase only supports two basic look up mechanisms based on the row key, and it does not provide a native mechanism for searching cell data. Therefore, an Elasticsearch-based secondary index is built to provide users with the ability to search metadata contents.

There are two procedures in this component according to Figure 3.10: data loading and index construction. In the data loading procedure, the data of indexed terms are extracted from HBase to Elasticsearch with the MapReduce program. In the Map stage, a number of `mappers` read cell values in HBase, extract data of indexed terms. After that, `mappers` generate a number of JSON objects based on extracted data values and commit a document insert request to the Elasticsearch cluster. In order to improve the speed of the data loading procedure, a temporary buffer is used to temporarily collect document insert requests committed by *mappers*. Once the commit conditions are satisfied, a set of document insert requests is bulk committed to the Elasticsearch cluster. Generally, the commit conditions are set to be the value of maximum cached commit requests.

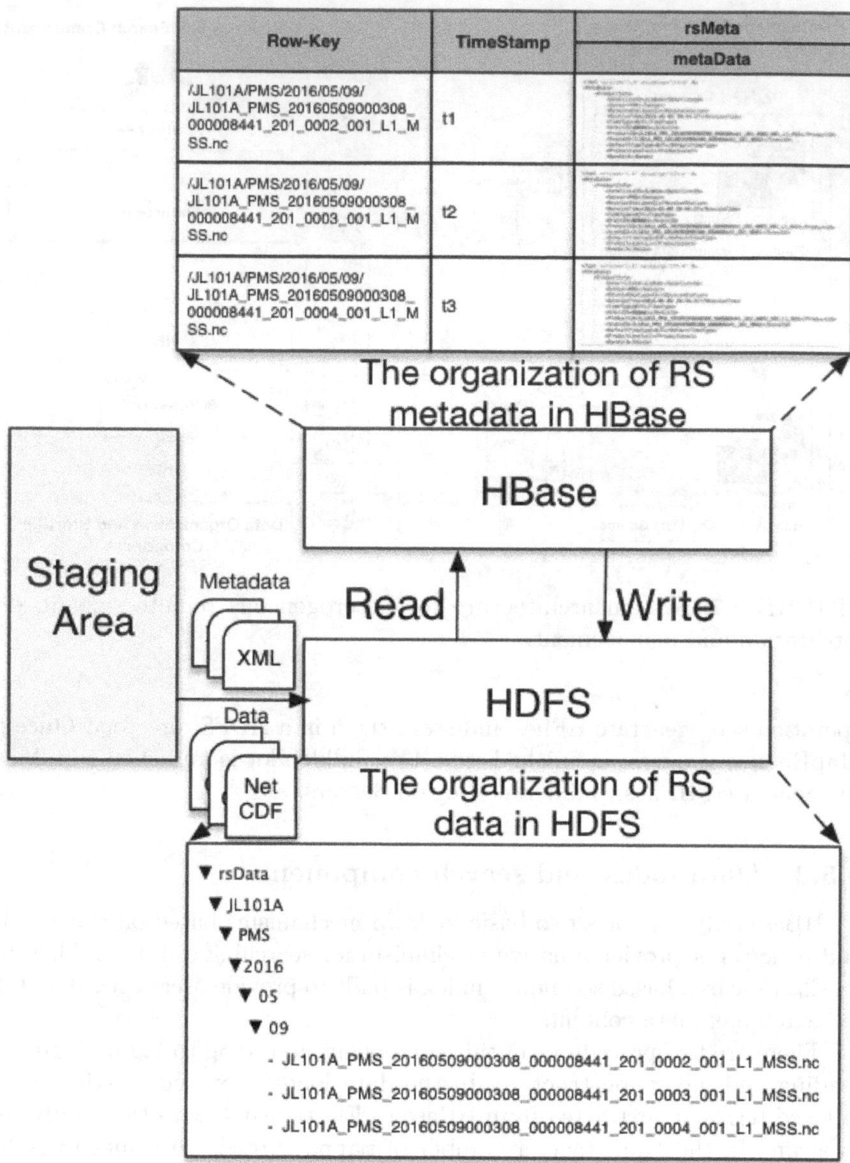

FIGURE 3.8: How remote sensing data and its metadata are organized in HDFS and HBase.

In the index construction procedure, a number of indexed terms need to be determined in advance. In the system, all the fields in Table 3.2 are selected as indexed terms besides centerLongitude and centerLatitude. The indexed terms are divided into the spatial data and the non-spatial data. The spatial data is composed of four geographic coordinates, and is stored as a geo_shape

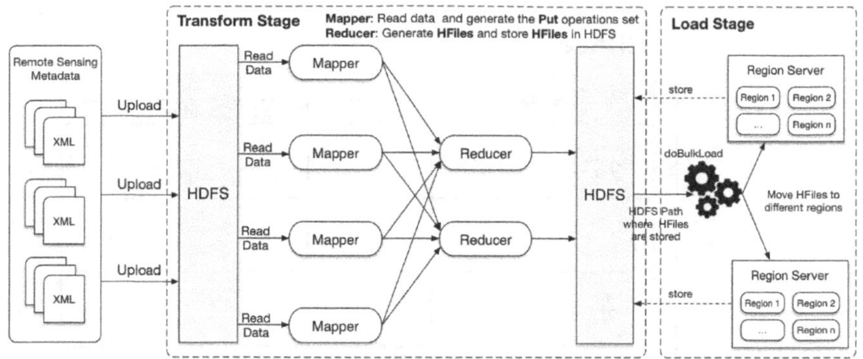

FIGURE 3.9: The procedure of the MapReduce-based bulkload technique to load remote sensing metadata from HDFS into HBase.

data type in Elasticsearch. The QuadPrefix technique is selected in the system to build the spatial index for remote sensing data. The QuadPrefix string as well as other non-spatial data are utilized to build the inverted index. By this means, the Elasticsearch cluster is able to provide users with different searching types, such as the spatial query and the text query.

3.6 Metadata Retrieval Experiments in a Parallel File System

Metadata retrieval experiments in a parallel file system were conducted in three Linux virtual machines. They were set up based on the OpenStack cloud computing framework. All of the three machines were developed with the SolrCloud environment, responsible for metadata index and retrieval. Each machine was mounted with a one-terabyte (TB) cloud drive so as to provide image storage space. In addition, all the machines in the main and sub centers were configured with 4 gigabytes (GBs) of RAM and 2 virtual processor cores.

3.6.1 LSI model-based metadata retrieval experiments in a parallel file system

In order to verify the retrieval efficiency for massive and multi-source remote sensing data in a parallel file system, we simulated about 15 million remote sensing metadata files. All of them are organized by the LSI model and imported into our metadata index and retrieval system SolrCloud. The total amount of our experimental metadata is shown in Table 3.2.

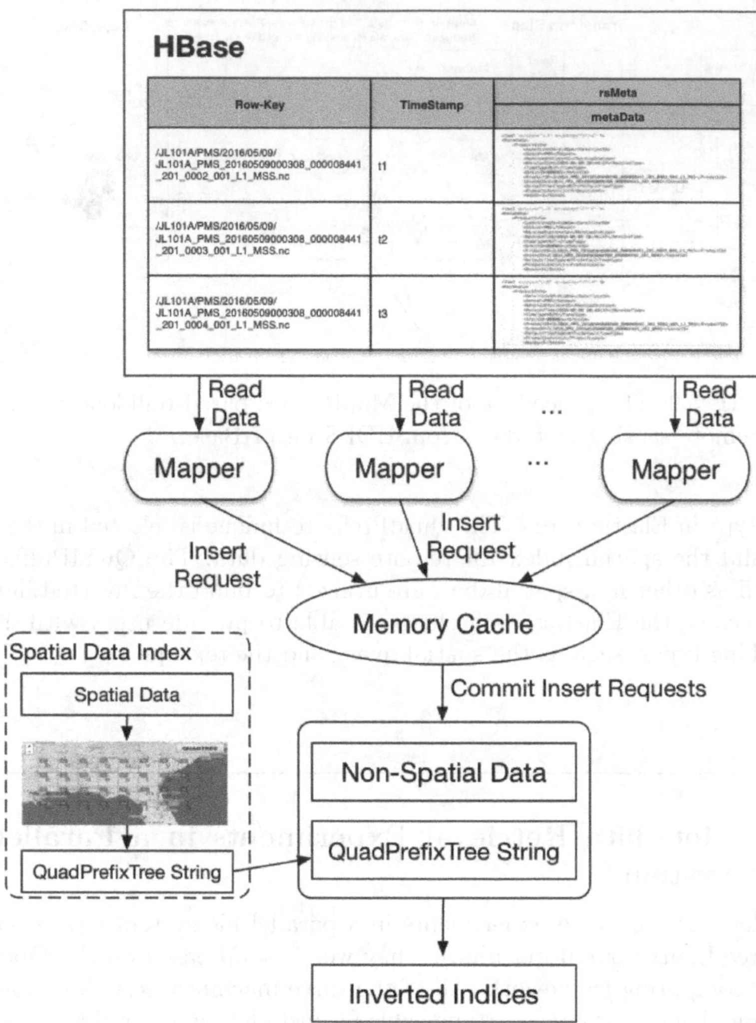

FIGURE 3.10: The procedure of the MapReduce-based bulkload technique load remote sensing metadata from HDFS into HBase.

TABLE 3.2: A summary of experimental remote sensing metadata.

Satellite	Data Type	Volume of Metadata	Metadata Format
Landsat 8	OLI_TIRS	896,981	HDF-EOS
HJ-1A	HSI	85,072	Customized XML
CEBERS-1/2	CCD	889,685	Customized XML
Landsat 7	ETM+	2,246,823	HDF-EOS
Landsat1-5	MSS	1,306,579	HDF-EOS
HJ-1A/B	CCD	2,210,352	Customized XML
Landsat 5	TM	2,351,899	HDF-EOS
FY-3A/B	VIRR	2,343,288	Customized HDF5-FY
Aster	L1T	2,951,298	HDF-EOS

In order to test the retrieval capabilities for different volumes of big data, the 15 million pieces of remote sensing metadata were copied and divided into six groups, and the volumes of the groups were 1 million, 3 million, 5.5 million, 7.5 million, 10 million, and 15 million. For the follow-up experiments, each group contained only 896,981 Landsat 8 OLI_TIRS pieces of metadata. In all the following experiments, we always set the platform and sensor parameters to Landsat 8 and OLI_TIRS, respectively, with only spatial and time parameters changing.

As for the spatial query parameters, there were several: parameters within 1 GeoSOT grid, 2 GeoSOT grids, and 4 GeoSOT grids. Therefore, the retrieval experiment of each group was divided into three subgroups. As for the time query parameters, in each subgroup of experiments, the query time frames were set to one day, one month, and six months, in order to verify the performance of our proposed method thoroughly. Furthermore, in order to exclude the influence of accidental factors, 20 retrievals were executed separately in each experiment and the average query time was the final result.

In addition, taking the GeoSOT code of the center point as the spatial identification of each remote sensing data is equivalent to converting polygon queries to point queries. It will improve the efficiency of data query, but the precision will be discounted. Therefore, a second filtering process, using longitude and latitude, should be made for the query results. Since the first query had ruled out the vast majority of irrelevant data, the second filtering process took little time. Hence, the efficiency of LSI model-based data query was very high. The search conditions of each group of experiment and time consumed are as shown in Table 3.3.

TABLE 3.3: The search conditions and time consumed of each retrieval.

Group	**Subgroup**	**Query Time Frames**		
Metadata Volume (Million)	**Spatial Parameters**	**1 Day**	**1 Month**	**6 Months**
1	1 GeoSOT Grid	133 ms	144 ms	145 ms
	2 GeoSOT Grids	139 ms	144 ms	151 ms
	4 GeoSOT Grids	151 ms	154 ms	155 ms
3	1 GeoSOT Grid	211 ms	213 ms	215 ms
	2 GeoSOT Grids	218 ms	224 ms	235 ms
	4 GeoSOT Grids	220 ms	239 ms	261 ms
5.5	1 GeoSOT Grid	310 ms	324 ms	325 ms
	2 GeoSOT Grids	340 ms	359 ms	375 ms
	4 GeoSOT Grids	365 ms	398 ms	421 ms
7.5	1 GeoSOT Grid	340 ms	350 ms	355 ms
	2 GeoSOT Grids	401 ms	405 ms	421 ms
	4 GeoSOT Grids	457 ms	476 ms	510 ms
10	1 GeoSOT Grid	480 ms	495 ms	525 ms
	2 GeoSOT Grids	566 ms	589 ms	603 ms
	4 GeoSOT Grids	650 ms	668 ms	691 ms
15	1 GeoSOT Grid	613 ms	655 ms	681 ms
	2 GeoSOT Grids	850 ms	856 ms	861 ms
	4 GeoSOT Grids	965 ms	994 ms	1110 ms

As can be seen in Table 3.3, in each group and subgroup, with the increase of query time frames, the time consumed showed an upward trend as a whole. However, the increase was not obvious. This type of situation could benefit from the inverted index of SolrCloud.

The small amount of time increment was mainly spent in the query results return process. As for the spatial query parameters changing, the time consumed within the 4-GeoSOT grid query was clearly greater than that within 1 GeoSOT grid, and this gap increased with the amount of metadata. This is perfectly understandable. The 4-GeoSOT grid query was the worst condition, and the comparison with the center point GeoSOT code of each remote sensing image should be made four times. However, within a 1 GeoSOT grid query, a one-time comparison would obviously be faster. Whether it is the spatial query parameters or query time frames that are changing, the retrieval times increase linearly with the increase of metadata volume. More specifically, the times for an increase rate below 10 million are a little shorter than those for an increase rate of 10 to 15 million.

3.6.2 Comparative experiments and analysis

In order to fully prove the superiority of our proposed LSI model-based metadata retrieval method, the following comparative experiments and analysis were carried out. Each type of comparative experiment contained six groups, and each group of experiments was carried out under the same data volumes and the same query parameters as the LSI model-based metadata retrieval experiments, using 20 average response time measurements [53].

3.6.2.1 Comparative experiments

(1) In order to show the advantages of our proposed LSI mode, the longitude and latitude were directly used to perform a full-text search, and other parameters were the same as in the LSI model-based experiments. For simplicity, the LSI model-based metadata retrieval method is simply referred to as SolrCloudLSI, and the longitude and latitude retrieval method is referred to as SolrCloudLatLon.

(2) In order to show the big data management and retrieval capabilities of SolrCloud, we built a single Solr node environment in a new virtual machine, with the same configuration as the SolrCloud nodes. The comparative experiment included two types: LSI model-based data retrieval, and the longitude- and latitude-based data retrieval on the single Solr node. The query parameters of the two types of experiments were the same as the LSI model-based data retrieval experiments. Similarly, the LSI model-based data retrieval on the single Solr node is referred to as SolrLSI, and the longitude- and latitude-based data retrieval on the single Solr node is referred to as SolrLatLon.

(3) In order to show the superiority of our proposed data management scheme with respect to other existing schemes, we chose HBase as the

comparison object. As a column-oriented key-value data store, HBase has been admired widely because of its lineage with Hadoop and HDFS [77, 78]. Therefore, LSI model-based data retrieval and the longitude- and latitude-based data retrieval experiments in HBase clusters were carried out. The cluster was provisioned with one NameNode and two DataNodes. The NameNode and DataNodes were configured in the same way as the SolrCloud cluster, 2 virtual processor cores and 4 GB of RAM. Hadoop 2.7.3, HBase 0.98.4 and Java 1.7.0 were installed on both the NameNode and the DataNodes. The query parameters and metadata volume of comparative experiments in the HBase cluster were the same as in the above experiments. Similarly, the LSI model-based data retrieval in the HBase cluster is referred to as HBaseLSI, and the longitude- and latitude-based data retrieval is referred to as HBaseLatLon.

The time consumptions of all comparative experiments are shown in Figure 3.11.

3.6.2.2 Results analysis

As can be seen in Figure 3.11, the following conclusions can be made.

Case 1: The spatial and time query parameters remained. In this case: (a) when the amount of metadata was less than 7.5 million items, the time consumption of the LSI model-based retrieval method was a little less than that of longitude- and latitude-based data retrieval; (b) with the increase of the metadata volume, the LSI model-based data retrieval was more efficient than the longitude- and latitude-based data retrieval; (c) when the amount of metadata was less than 5.5 million items, the time consumption of LSI model-based metadata retrieval on a single Solr node was not very different from that of SolrCloud; (d) when the metadata volume increased, the retrieval speed differences between SolrCloud and Solr became larger; (e) as for the longitude- and latitude-based data retrieval on the single Solr node, its retrieval speed was much slower than that of our proposed metadata retrieval program; and (f) although the query time increased little with the increase of metadata volume in the HBase cluster, it was still larger than that of the LSI model-based method. This may be because HBase has to manually scan the entire database to get results if we try to "filter" based on a "component" of the key or any of the values [79]. SolrCloud, on the other hand, with its inverted index, can handle queries on any of the fields in any combination, and can simply blaze them fast.

Case 2: The spatial query parameters remained but time frames changed. In this case: (a) with the increase of query time frames, the time consumed showed an upward trend as a whole, but this was not obvious, not only for SolrCloud but also in the Solr single node—this type of situation could benefit from the inverted index of SolrCloud and Solr; and (b) the query time increased little with the increase of query time frames in the HBase cluster.

Case 3: The time frames remained but spatial query parameters changed. In this case: (a) the time consumption increased with the increase of query

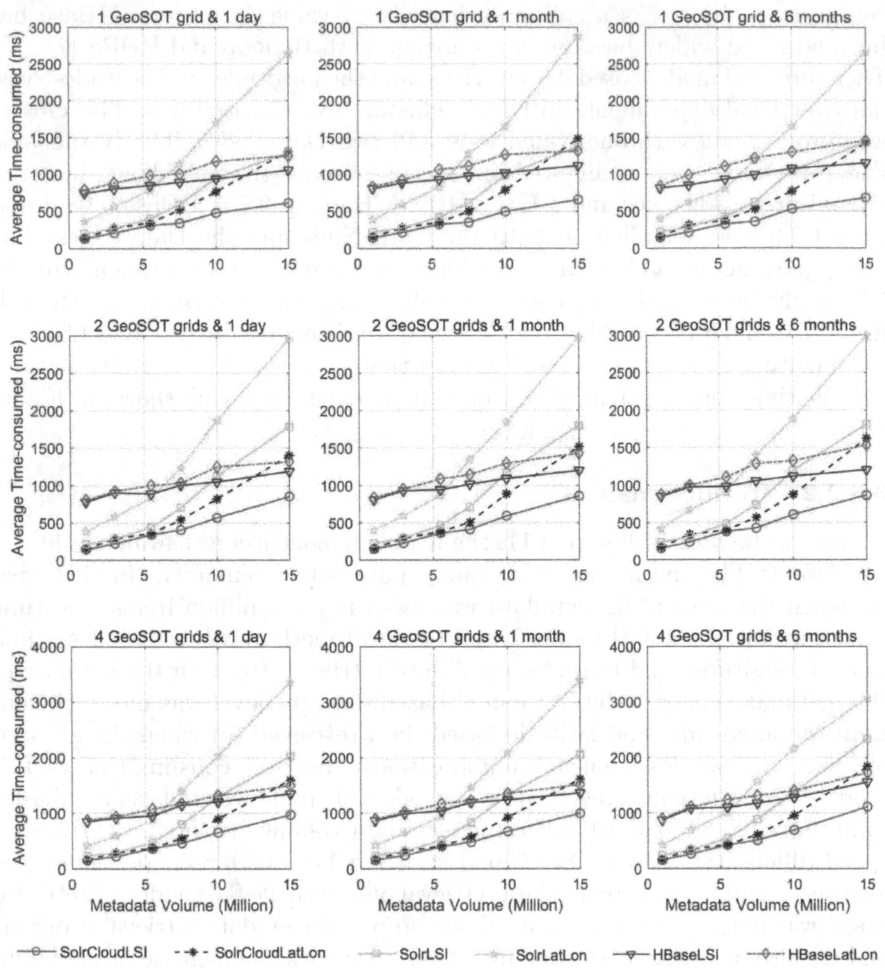

FIGURE 3.11: The results of all comparative experiments.

spatial extent, regardless of using the LSI model-based metadata retrieval method or the longitude- and latitude-based data retrieval method; and (b) for both the SolrCloud/Solr and the HBase cluster, the query time growth rate of the LSI model-based metadata retrieval method was greater than that of the longitude- and latitude-based data retrieval method. This may be because the comparison increased with the GeoSOT grid number increase. However, such a small flaw still did not affect the overall query efficiency of the LSI model-based data retrieval method.

In short, the results of all the comparative experiments proved the superiority of our proposed data retrieval program.

3.7 Metadata Retrieval Experiments in the Hadoop Ecosystem

Experiments are conducted on a cluster with six physical nodes. One node is the master node and the others are slave nodes. The master node is equipped with 2 × 6-core Intel Xeon@CPUs at 2.60 GHz, 16 GB of RAM and two 2 TB SATAs. The slave nodes are equipped with 2 × 6-core Intel Xeon@CPUs at 2.60 GHz, 64 GB of RAM, one 300 GB SAS and two 2 TB SATAs. Note that all disks in each node are set up with RAID 0. Two clusters are built in our experiments: a Hadoop cluster and an Elasticsearch cluster. (1) The Hadoop cluster contains six physical nodes. The master node plays the role of the namenode in Hadoop and the HMaster in HBase, and the other five slave nodes act as Hadoop datanodes and HBase region servers. Apache Hadoop 2.7.5, Apache Zookeeper 3.4.10 and Apache HBase 1.4.0 are adopted in our experiments. Besides, each block in HDFS has three replicas and the block size is set to be 128 MB. (2) The Elasticsearch cluster is running on the SSD disk of five slave physical nodes. The Elasticsearch version we adopt in experiments is 5.6.5. An index called `rsindex` is created to index terms selected from remote sensing metadata. The index has five shards, each of which has one primary shard and one replica shard. Additionally, Apache OODT with version 1.2.0 is deployed in the master node to integrate distributed remote sensing data and its metadata.

The experimental remote sensing metadata are freely downloaded from USGS (`https://earthexplorer.usgs.gov/`). More details about the metadata are listed in Table 3.4. The downloaded remote sensing data need to be pre-processed based on the ISO 19115-2:2009 metadata template.

TABLE 3.4: Descriptions of experimental remote sensing metadata downloaded from USGS.

Satellite	Sensor	Volume of Metadata
HJ1A	CCD1	2,033
HJ1B	CCD1	947
EO-1	ALI	86,118
	Hyperion	85,073
NPP	VIIRS	919,987
Landsat3	MSS	524,209
Landsat5	MSS	795,164
Landsat7	ETM	2,422,349
LandSat8	OLI_TIRS	1,167,511
TERRA	ASTER	3,190,775

3.7.1 Time comparisons of storing metadata in HBase

In order to show the efficiency of the proposed MapReduce-based bulkload method in loading metadata from HDFS to HBase, a number of comparative experiments are conducted. Each comparative experiment is carried out under the same data volumes. The technique of loading metadata from a local file system to HBase with native HBase APIs is referred as noBulkLoad. The technique of loading metadata from HDFS to HBase with the MapReduce-based bulkload technique is referred as bulkLoad. In the system, the total time of the bulkload is the sum of the time loading metadata from the local file system to HDFS and the time loading metadata from HDFS to HBase. The time consumed in the data loading procedure with noBulkLoad technique is denoted as $T_{noBulkLoad}$. The time consumed in the data loading procedure with bulkLoad technique is denoted as $T_{bulkLoad}$. The speedup ratio is computed by $T_{noBulkLoad}/T_{bulkLoad}p$. Comparison results are listed in Table 3.5. The time consumed by loading metadata into HBase is significantly decreased with our designed MapReduce-based bulkload mechanism. The speedup ratio is up to 87.

TABLE 3.5: Comparisons for storing metadata into HBase.

Volume of Metadata	$T_{noBulkLoad}$ (ms)	$T_{bulkLoad}$ (ms)	Speedup Ratio
1,000,000	5,618,394	64,239	87.46
2,000,000	9,806,346	120,571	81.33
3,000,000	13,981,113	179,990	77.68
4,000,000	18,146,389	235,938	76.91
5,000,000	22,287,981	293,886	75.84
6,000,000	26,434,366	347,969	75.97
7,000,000	30,582,689	407,421	75.06
8,000,000	34,726,448	460,642	75.39
9,000,000	38,870,983	521,324	75.56

3.7.2 Time comparisons of loading metadata from HBase to Elasticsearch

In order to show the efficiency of the proposed MapReduce-based bulk load method in loading metadata contents from HBase to Elasticsearch, a number of comparative experiments are conducted. Each comparative experiment is carried out under the same data volume. The technique of loading metadata contents from HBase to Elasticsearch with no cache mechanism is referred as bulkLoad. The technique of loading metadata contents from HBase to Elasticsearch is referred as bulkLoadWithCache. The time consumed in the data loading procedure with bulkLoad technique is denoted as $T_{bulkLoad}$, and the time consumed in the data loading procedure with bulkLoadWithCache is denoted as $T_{bulkLoadWithCache}$. The speedup ratio is computed by $T_{noBulkLoad}/T_{bulkLoad}p$. In addition, the value of maximum cached commit requests is set to 2000 in our experiments. Table 3.6 presents experimental results. The utilization of a temporary buffer alleviates the time consumed by extracting metadata contents from HBase to Elasticsearch.

TABLE 3.6: Comparisons for loading metadata from HBase to Elasticsearch.

Volume of Metadata	$T_{bulkload}$ (ms)	$T_{bulkLoadWithCache}$ (ms)	Speedup Ratio
1,000,000	22,381,389	3,471,814	6.45
2,000,000	44,492,357	6,674,109	6.67
3,000,000	66,106,287	9,970,543	6.63
4,000,000	88,574,159	13,642,927	6.49
5,000,000	114,217,929	17,483,857	6.53
6,000,000	135,707,799	20,380,396	6.66
7,000,000	165,344,836	25,076,789	6.59
8,000,000	184,007,335	27,683,614	6.65
9,000,000	215,692,456	33,081,665	6.52

3.8 Conclusions

In view of the current issues of remote sensing data management, this chapter proposed a SolrCloud-based data management framework. Specifically, for efficient retrieval problems of integrated massive data, we proposed the LSI model-based data organization approach, and took SolrCloud to realize the distributed index and quick retrieval of metadata. In addition, in order to verify the availability of our proposed program, a series of distributed data retrieval and comparative experiments were carried out. The results showed that our proposed distributed data management program was effective and provided superior capabilities. In particular, the LSI model-based data organization and the SolrCloud-based distributed indexing schema could effectively improve the efficiency of massive data retrieval.

In addition, two popular big data systems Hadoop and HBase are selected to store remote sensing data gathered from distributed data sources. Additionally, an Elasticsearch-based secondary indexing technique is designed to support the searching of remote sensing metadata contents.

However, the GeoSOT code length of each remote sensing image was calculated according to the image swath. This calculation is easy, and the obtained GeoSOT code is not very long. These relatively short GeoSOT codes could not bring a heavy query burden. However, despite this disadvantage, these relatively short GeoSOT codes, to a certain degree, have reduced query accuracy. Thus, future work will be focused on exploring a suitable GeoSOT code length calculation method, such as introducing the feedback control theory to calculate GeoSOT code length of each type of remote sensing image, so that neither the query efficiency nor accuracy will be affected.

Chapter 4

High Performance Remote Sensing Data Processing in a Cloud Computing Environment

4.1 Introduction

With the remarkable advances in high-resolution Earth Observation (EO), we are witnessing an explosive growth in the volume and also velocity of Remote Sensing (RS) data. The latest-generation space-borne sensors are capable of generating continuous streams of observation data at a growing rate of several gigabytes per second ([80]) almost every hour, every day, every year. The global archived observation data probably exceed one exabyte according to the statistics of an OGC report ([81]). The volume of RS data acquired by a regular satellite data center is dramatically increasing by several terabytes per day, especially for the high-resolution missions ([82]), while, the high-resolution satellites, namely indicating higher spatial, higher spectral and higher temporal resolution of data, which would inevitably give rise to the higher dimensionality nature of pixels. Coupled with the diversity in the present and upcoming sensors, RS data are commonly regarded as *"Big RS Data"*or *"Big Earth Observation Data"*, not merely in data volume, but also in terms of the complexity of data.

The proliferation of "RS Big Data" is revolutionizing the way RS data are processed, analyzed and interpreted as knowledge ([83]). In large-scale RS applications, regional or even global covered multi-spectral and multi-temporal RS datasets are exploited for processing, so as to meet the rising demands for more accurate and up-to-date information. A continent-scale forest mapping normally involves processing terabytes of multi-dimensional RS datasets for available forest information ([84]). Moreover, large-scale applications are also exploiting multi-source RS datasets for processing so as to compensate for the limitation of a single sensor. Accordingly, not only the significant data volume, but the increasing complexity of data has also become the vital issue. Particularly, many time-critical RS applications even demand real-time or near real-time processing capacities ([85][86]). Some relevant examples are large debris flow investigation ([87], flood hazard management ([88]) and large ocean oil spills surveillance ([89][90])). Generally, these large-scale data processing problems in RS applications ([82][91][92]) with high QoS requirements are typically regarded as both compute-intensive and data-intensive. Likewise, the innovative analyses and high QoS (Quality of Service) requirements are driving the renewal of traditional RS data processing systems. The timely processing of tremendous multi-dimensional RS data has introduced unprecedented computational requirements, which is far beyond the capability that conventional instruments could satisfy. Employing a cluster-based HPC (High-Performance Computing) paradigm in RS applications turns out to be the most widespread yet effective approach ([93][94][95][96][97]). Both NASA's NEX system ([83] for global processing and InforTerra's "Pixel Factory" ([98]) for massive imagery auto-processing adopt cluster-based platforms for QoS optimization.

However, despite the enormous computational capacities, cluster platforms that are not data-intensive optimized are still challenged with huge data analysis and intensive data I/O. The mainstream multi-core clusters are characterized with multilevel hierarchy and increasing scale. These HPC systems are almost out of reach for non-experts of HPC, since the easy programming on MPI-enabled (Message Massing Interface) HPC platforms is anything but easy. Moreover, the prevalent on-line processing needs are seldom met, in that, there lacks an easy-to-use way to serve end users the massive RS data processing capabilities in large-scale HPC environment ubiquitously. However, diverse RS data processing typically follows a multi-stage on-the-flow processing. The on-demand processing also means the ability to customize and serve dynamic processing workflows, instead of the predefined static ones, while, on-demand provision of resources will result in unpredictable and volatile requirements of large-scale computing resources. A substantial investment for sustaining system upgrade and scale-out would be essential. In addition, the building and maintenance of such platforms is remarkably complicated and expensive.

Cloud computing ([99]) provides scientists with a revolutionary paradigm of utilizing computing infrastructure and applications. By virtue of virtualization, the computing resources and various algorithms could be accommodated and delivered as ubiquitous services on-demand according to the application requirements. The Cloud paradigm has also been widely adopted in large-scale RS applications, such as the Matsu project ([100]) for cloud-based flood assessment. Currently, Clouds are rapidly joining HPC systems like clusters as variable scientific platforms ([101]). Scientists could easily customize their HPC environment and access huge computing infrastructures in the Cloud. However, compared to conventional HPC systems or even supercomputers, the Clouds are not QoS-optimized large-scale platforms. Moreover, differing from the traditional Cloud, these Datacenter Clouds deployed with data-intensive RS applications should facilitate massive RS data processing and intensive data I/O.

To efficiently address the aforementioned issues, we propose *pipsCloud*, a cloud-enabled High-Performance RS data processing system for large-scale RS applications. The main contribution of it is that it incorporates a Cloud computing paradigm with cluster-based HPC systems in an attempt to address the issues from a system architecture point of view. Firstly, by adopting application-aware data layout optimized data management and Hilbert R^+ tree based data indexing, the RS big data including imageries, interim data and products could be efficiently managed and accessed by users. By means of virtualization and bare-metal (BM) provisioning ([102]), not only virtual machines, but also bare-metal machines with less performance penalty are deployed on-demand for easy scale up and out. Moreover, the generic parallel programing skeletons are also employed for easy programming of efficient MPI-enabled RS applications. Following this way, the cloud-enabled virtual HPC environment for RS big data processing is also dynamically encapsulated and delivered as on-line services. Meanwhile, benefiting from a dynamic scientific workflow technique, *pipsCloud* offers the ability to easily customize collaborative processing workflows for large-scale RS applications.

The rest is organized as follows. Section 4.2 reviews some related works, and Section 4.3 discuss the challenges lying in the building and enabling a high performance cloud system for data-intensive RS data processing. Section 4.4 demonstrates the design and implementation of the *pipsCloud* from the system level point view. Then Section 4.5 discusses the experimental validation and analysis of the *pipsCloud*. Finally Section 4.6 concludes this chapter.

4.2 High Performance Computing for RS Big Data: State of the Art

Each solution has its pros and cons. In this section, we comparatively review current dominant system architectures regularly adopted in the context of RS data processing, both cluster-based HPC platforms and Clouds. Firstly, in Section 4.2.1, we go deep into the incorporation of multi-core cluster HPC structure with RS data processing systems and applications. Then, in Section 4.2.2, we introduce some new attempt to enable large-scale RS applications by taking advantage of a Cloud computing paradigm.

4.2.1 Cluster computing for RS data processing

As increasing numbers of improved sensor instruments are incorporated with satellites for Earth Observation, we have been encountering an era of "RS Big Data." Meanwhile, the urgent demands for large-scale remote sensing problems with boosted computation requirements ([83]) have also fostered the widespread applying of multi-core clusters. The first shot goes to the NEX system ([83]) for global RS applications built by NASA on a cluster platform with 16 computer in the middle of 1990s. The "Pixel Factory" system ([98]) of InforTerra employed a cluster-based HPC platform for massive RS data auto-processing, especially Ortho-rectification. These HPC platforms are also employed in the acceleration of hyperspectral imagery analysis ([103]). It is worth noting that the 10,240-CPU Columbia supercomputer[1] equipped with InfiniBand network has been exploited for remote sensing applications by NASA.

Several traditional parallel paradigms are commonly accepted for these multi-level hierarchy featured cluster systems. OpenMPparadigm is designed for shared-memory, MPI is adopted within or across nodes, and the MPI+OpenMP hybrid paradigm ([104]) is employed for exploiting multilevels of parallelism. Recently, great efforts have been made in the incorporation of an MPI-enabled paradigm with remote sensing data processing in the large scale scenarios. Some

[1]Columbia Supercomputer at NASA Ames Research Center, http://www.nas.nasa.gov/Resources/Systems/columbia.html

related works with Plaza et al. presented parallel processing algorithms for hyperspectral imageries ([105]), Zhao et al. ([106]) implemented soil moisture estimation in parallel on a PC cluster, as well as MPI-enabled implementing of image mosaicking ([107], fusion ([108]) and band registration ([109]). Obviously, benefiting from the efforts and developments conducted in HPC platforms, plenty of RS applications have enhanced their computational performance in a significant way ([83]).

However, in spite of the elegant performance acceleration has achieved, it is still anything but easy for non-experts to employ the cluster-based HPC paradigm. Firstly, the programming, deploying as well as implementing of parallel RS algorithms on an MPI-enabled cluster are rather difficult and error-prone ([110]). Secondly, HPC systems are not optimized for data-intensive computing especially. The loading, managing and communication of massive multi-dimensional RS data on the distributed multilevel memory hierarchy of an HPC system would be rather challenging. Some emerging PGAS ([111]) typed approaches offer global but partitioned memory address spaces across nodes, like UPC ([112], Chapel ([113]) and X10 ([114]). The on-going DASH project[2] is developing Hierarchical Arrays (HA) for hierarchical locality. Thirdly, the relatively limited resources in HPC systems could not be easily scaled to meet the on-demand resource needs of diverse RS applications. For affordable large-scale computing resources, substantial upfront investment and sustaining scaling up would be inevitable but also rather expensive. In addition, cluster-base HPC systems lack an easy and convenient way of utilizing high performance data processing resources and applications, not to mention the on-demand customizing of computing resources and processing workflows.

4.2.2 Cloud computing for RS data processing

The Cloud has emerged as a promising new approach for ad-hoc parallel processing, in the Big Data era[115]. It is capable of accommodating variable large-scale platforms for different research disciplines with elastic system scaling. Benefiting from virtualization, not only computing resources, but also software could be dynamically provisioned as ubiquitous services best suited to the needs of the given applications. Compared to the MPI-enabled cluster systems, the cloud paradigm provides computing resources in a more easy-to-use and convenient way – a service-oriented way.

The advent of the Cloud has also empowered remote sensing and relevant applications. Matsu ([100], the on-going research project of NASA for on-demand flood prediction and assessment with RS data adopts an Eucalyptus-based[116]) distributed cloud infrastructure with over 300 cores. GENESI-DEC[3], a project of the Ground European Network for Earth Science Interoperations – Digital Earth Communities. It employs a large and distributed cloud infrastructure to

[2]DASH project: under DFG programme "Software for Exascale Computing – SPPEXA", http://www.dash-project.org

[3]GENESI-DEC, http://www.genesi-dec.eu

allow worldwide data access, produce and share services seamlessly. With the virtual organization approach, the Digital Earth Communities could lay their joint effort for addressing global challenges, such as biodiversity, climate change and pollution.The ESA (European Space Agency) G-POD[4], a project to offer on-demand processing for Earth observation data was initially constructed with GRID. Subsequently, the Terradue cloud infrastructure was selected to enhance G-POD for resources provisioning ([83]).

Great efforts have been made in the employing of cloud computing in the context of remote sensing data processing, both in terms of programming models and resource provisioning.

4.2.2.1 Programming models for big data

Several optional distributed programming models[115] are prevalently employed for processing large data sets in the cloud environment, like MapReduce ([117]) and Dryad[5], where, MapReduce is the most widely accepted model for distributed computing in Cloud environment. By using "Map" and "Reduce" operations, some applications could be easily implemented in parallel without concerning data splitting and any other system related details. With the growing interest in Cloud computing, it has been greatly employed in RS data processing scenarios. Lin et al. ([118]) proposed a service integration model for GIS implemented with MapReduce, B. Li et al. ([119]) employed MapReduce for parallel ISODATA clustering. Based on the Hadoop MapReduce framework, Almeer ([120]) built an experimental 112-core high-performance cloud system at the University of Qatar for parallel RS data analysis.

Despite the simple but elegant programming feature, MapReduce is not a fits-all model. In the context of large-scale RS applications, we have data with higher dimensionality, algorithms with higher complexity as well as specific dependences between computation and data. In these scenarios, the simple data partition of MapReduce with no idea of actual applications would not always work for acceleration, not to mention the applications with numerical computing issues.

4.2.2.2 Resource management and provisioning

Essentially, on-demand resource managing and provisioning are foremost in the cloud computing environment. Several choices of open-source cloud solutions are available to accommodate computing infrastructure as a service for viable computing. Among several solutions, such as OpenStack ([121], OpenCloud[6], Eucalyptus ([116]) and OpenNebula ([122], OpenStack is the most widely accepted and promising one. Basically, in recent Clouds, on-demand resource allocation and flexible management are built on the basis of virtualization. Many available choices of hypervisors for Server virtualization in current

[4]Grid Processing on Demand, https://gpod.eo.esa.int

[5]Dryad, http://research.microsoft.com/en-us/projects/dryad/

[6]OpenCloud, https://www.opencloud.com

open cloud platforms are Xen hypervisor[123], Kernel-based Virtual Machine (KVM) ([124]) as well as VMWare ([125]). By management of Provisioning of Virtual Machines(VMs), hypervisors could easily scale up and down to provide a large-scale platform with a great number of VMs. Likewise, the network virtualization concept has also emerged. Yi-Man ([126]) proposed Virt-IB for InfiniBand virtualization on KVM for higher bandwidth and lower latency.

The virtualization approaches normally deploy multiple VMs instances on a single physical machine (PM) for better resource utilization. However, virtualization and hypervisor middleware would inevitably introduce an extra performance penalty. Recently, Varrette et. al. ([127]) has demonstrated the substantial performance impact and even a poor power efficiency when facing HPC-type applications, especially large-scale RS applications. As a result, whether the VMs in the Cloud suit as a desirable HPC environment is still unclear.

For performance sake, native hypervisors (compared to hosted hypervisors) that capable of bare-metal provisioning is another option to extend current cloud platforms. This kind of hypervisors implement directly on the hardware of hosts and take control of them. There are several bare-metal provisioning hypervisors (examples are xCAT[7], Perceus[8]). Xie ([102]) has extended OpenStack to support bare-metal provisioning using xCAT, so as to serve both VMs and bare-metal machines in a cloud platform.

4.3 Requirements and Challenges: RSCloud for RS Big Data

The Cloud computing paradigm has empowered RS data processing and makes it more possible than ever ([128]). Unlike conventional ways of processing that are done by standalone server or software, the cloud-based RS data processing is enabled with a revolutionary promise of unlimited computing resources.

Despite the advantages we could explore in the cloud computing paradigm, there remain some obstacles to cloud adoption in remote sensing. As we look into the future needs of large-scale RS exploration and applications, it is clear that the distributed "data deluge" and computing scales will continue to explode and even drive the evolution of processing platforms. Naturally, the tremendous RS data and RS algorithms may actually be distributed across multiple data centers located in different geographic places crossing organizational or even national boundaries. In this scenario, the efficient storing, managing as well as sharing and accessing of these distributed RS data at

[7]xCAT (Extreme Cloud Administration Toolkit), http://en.wikipedia.org/wiki/XCAT
[8]perceus, http://www.perceus.org

such an extreme volume and data complexity would be anything but easy. Likewise, large-scale RS applications and environmental research might always be in demand for collaborative workflow processing across data centers in a distributed environment by network. To allow pervasive and convenient service for domain users on-line in such a heterogeneous distributed environment, a one-stop service oriented user interface could be of vital importance. Moreover, different types of domain users as well as varieties of RS applications with different processing scales would normally give rise to diverse requirements of a cloud service model. The examples of these service models are RS data subscription and virtual data storage service, virtual cluster-base HPC platform service or even virtual processing system service for RS.

To meet the above challenging demands, a cloud-enabled HPC framework especially designed for the remote sensing era, namely RSCloud, is quite critical. Firstly, it should be capable of managing and sharing massive distributed RS data. Accordingly, RS big data across data centers can be easily accessed and subscribed on-demand as a virtualized data catalog or storage for global sharing. Secondly, it should also be able to offer an RS-specific HPC environment with not only elastic computing and storage capacities as well as RS-oriented programming and runtime environment so as to meet the on-demand need of applications. Thirdly, the dynamic workflow processing is also essential for addressing global RS problems which require the collaboration of multiple scientists or organizations. In addition, the one-stop service of the cloud portal is also important, where a wide variety of resources, and even on-line RS data processing could be easily accommodated through one-stop accessing. Last but not least, are different service models of a multi-tenant environment, where different types of virtual HPC environment and RS processing systems can be abstracted and served on-demand. Loosely speaking, RSCloud is a cloud-assisted platform for RS, which not only facilitates elastic provisioning of computing and storage resources, but also allows multi-tenant users on-demand access, sharing and a collaborative process of massive distributed RS data in different service models.

4.4 *pipsCloud*: High Performance Remote Sensing Clouds

To properly address the above issues, we propose *pipsCloud*, a high-performance RS data processing system for large-scale RS applications in the cloud platform. It provides a more efficient and easy-to-use approach to serve high-performance RS data processing capacity on-demand, and also QoS optimization for the data-intensive issues.

4.4.1 The system architecture of *pipsCloud*

As illustrated in Figure 4.1, *pipsCloud* adopts a multi-level system architecture. From bottom to top it is respectively physical resources, cloud framework, VE-RS, VS-RS, data management and cloud portal. The cloud framework manages physical resources to offer Iaas (Infrastructure as a Service) by virtue of OpenStack. Based on the cloud framework, the VE-RS offers a virtual HPC cluster environment as a service and VS-RS provides a cloud-enabled virtual RS big data processing system for on-line large-scale RS data processing, while the management, indexing and sharing of RS big data are also served as Daas (Data as a service).

Cloud Framework employs the most popular but successful open source project OpenStack to form the basic cloud architecture. However, OpenStack mostly only offers virtual machines (VMs) through virtualization technologies. These VMs are run and managed by hypervisors, such as KVM or Xen. Despite the excellent scalability, the performance penalty of virtualization is inevitable. To support the HPC cluster environment in the Cloud, pipsCloud adopts a bare-metal machine provisioning approach which extends OpenStack with a bare-metal hypervisor named xCAT. Following this way, both VMs and bare-metal machines could be scheduled by nova-scheduler and accommodated to users subject to application needs.

VE-RS, namely an RS-specific cluster environment with data-intensive optimization, which also provided as a VE-RS service based on the OpenStack enabled cloud framework. By means of auto-configuration tools like AppScale, VE-RS could build a virtual HPC cluster with Torque task scheduler and Ganglia for monitoring on top of the cloud framework. Then varieties of RS softwares could be customized and automatically deployed on this virtual cluster with SlatStack[9]. Moreover, a generic parallel skeleton together with a distributed RS data structure with fine-designed data layout control are offered for easy but productive programming of large-scale RS applications on an MPI-enabled cluster.

VS-RS, a virtual processing system that is built on top of VE-RS is served as an on-line service especially for large-scale RS data processing. A VS-RS not only provides RS data products processing services, but also offers a VS-RS processing system as a service to provide processing workflow customization. By virtue of the Kepler scientific workflow engine, VS-RS could offer dynamic workflow processing and also on-demand workflow customization. The thing that is worth noting is that enabled by Kelper, the complex workflow could also be built among clouds or different data centers with web services. Moreover, the RS algorithm depository together with the RS workflow depository are employed in VS-RS for workflow customization, interpreting and implementing. Besides, order management as well as system monitoring and management are also equipped in VS-RS to enable on-line processing and management.

---

[9]saltstack: https://docs.saltstack.com/en/latest/

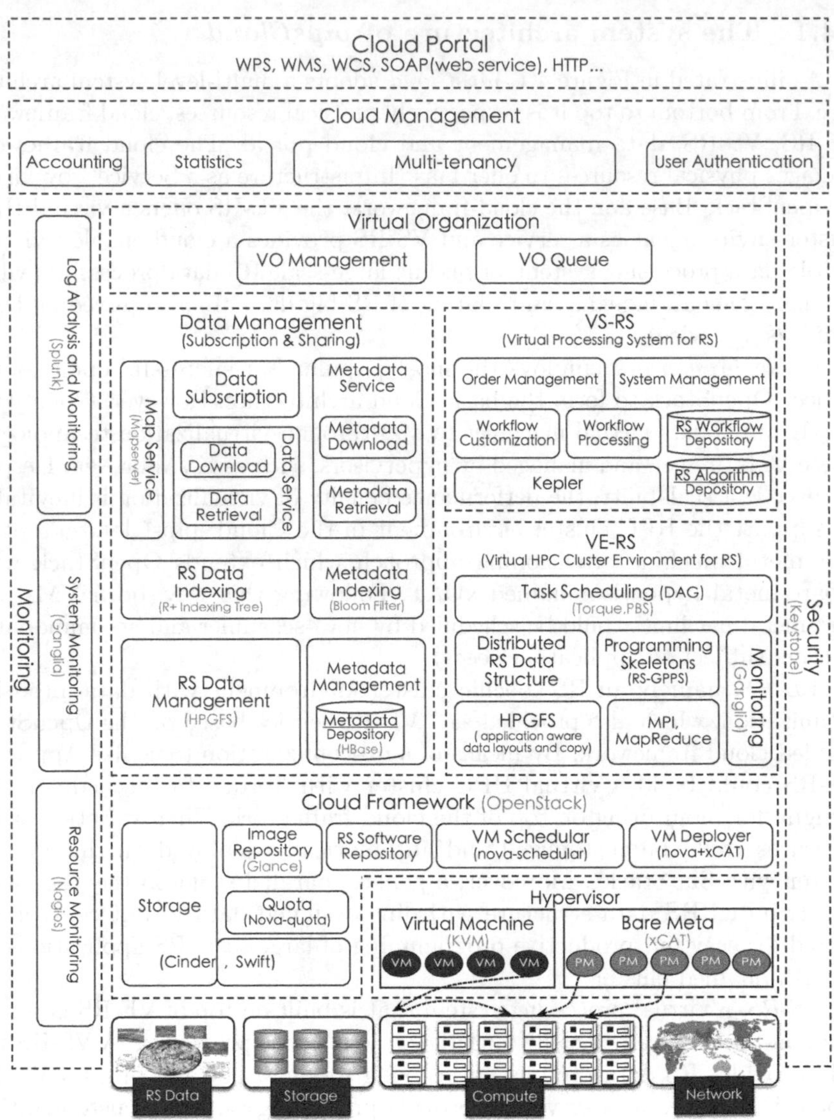

FIGURE 4.1: The system architecture of *pipsCloud*.

RS Data Management is a novel and efficient way of managing and sharing RS big data on top of the cloud framework. It adopts unbounded cloud storage enabled by Swift to store these varieties of data and serve them in a RS data as a service manner. HPGFS the distributed file system for an RS data object is used for managing enormous unstructured RS data, like RS imageries, RS data products and interim data, while the structured RS metadata are managed by a NoSQL database – HBase[129]. For quick retrieval from varieties of massive

RS data, a Hilbert R^+ tree together with an in-memory hot data cache policy are used for indexing acceleration. Last but not least, the thesis-based data subscription with virtual data catalog and thesis-based data push are also put forward as a novel means of data sharing.

Cloud Management and Portal manages the whole *pipsCloud* platform, including system monitoring, user authentication, multi-tenancy management as well as statistics and accounting. While in the web portal of *pipsCloud*, the RS data management, RS data processing capabilities as well as on-demand RS workflow processing are all encapsulated as OGS web services interface standards, such as WPS (Web Proccessing Service), WCS (Web Coverage Service) and WMS (Web Map Service).

4.4.2 RS data management and sharing

Efficient RS data management and sharing are paramount especially in the context of large-scale RS data processing. The managing of RS big data is not only limited to unstructured multi-source RS imageries, but also varieties of RS data products, interim data generated during processing, as well as structured metadata. As is mentioned above, HPGFS with application-aware data layout optimization is adopted for managing unstructured RS data, while the RS metadata are stored in HBase for query. These unstructured RS data are organized in a GeoSOT global subdivision grid, and each data block inside these data are arranged in Hilbert order and encoded with a synthetic GeoSOT-Hilbert code combining GeoSOT code with Hilbert value. The GeoSOT-Hilbert together with the info of data blocks are stored in the column family of the metadata in HBase for indexing. For a quick retrieval from varieties of massive RS data, a Hilbert R^+ tree with GeoSOT[10] [130] global subdivision is employed for indexing optimization. For further indexing acceleration, a hot-data cache policy is adopted to cache "hot" RS imageries and metadata into the Redis[131][132] in-memory database and offer hash table indexing. The most important thing that is worth mention is the easy but novel means of RS data sharing – thesis-based RS data subscription and data push through virtual data catalog mounting as local.

The runtime implementing of RS data management and sharing is demonstrated in Figure 4.2. First, pipsCloud interprets the data requests, and checks the user authentication. Second, it conducts a quick search in the "hot" data cache on the Redis in-memory database with the hash table index; if the cache hits then it returns data. Third, it searches the required data in the Hilbert R^+ tree for the unique Hilbert-GeoSOT code. Fourth, it uses the Hilbert-GeoSOT code of the found data to locate the metadata entry in HBase, or locate the physical URL of the data for accessing. Fifth, for a subscription request, it re-organizes these data entries to form a virtual mirror and mount it to user's

[10]GeoSOT: Geographical Coordinate Subdividing Grid with One Dimension Integer Coding Tree

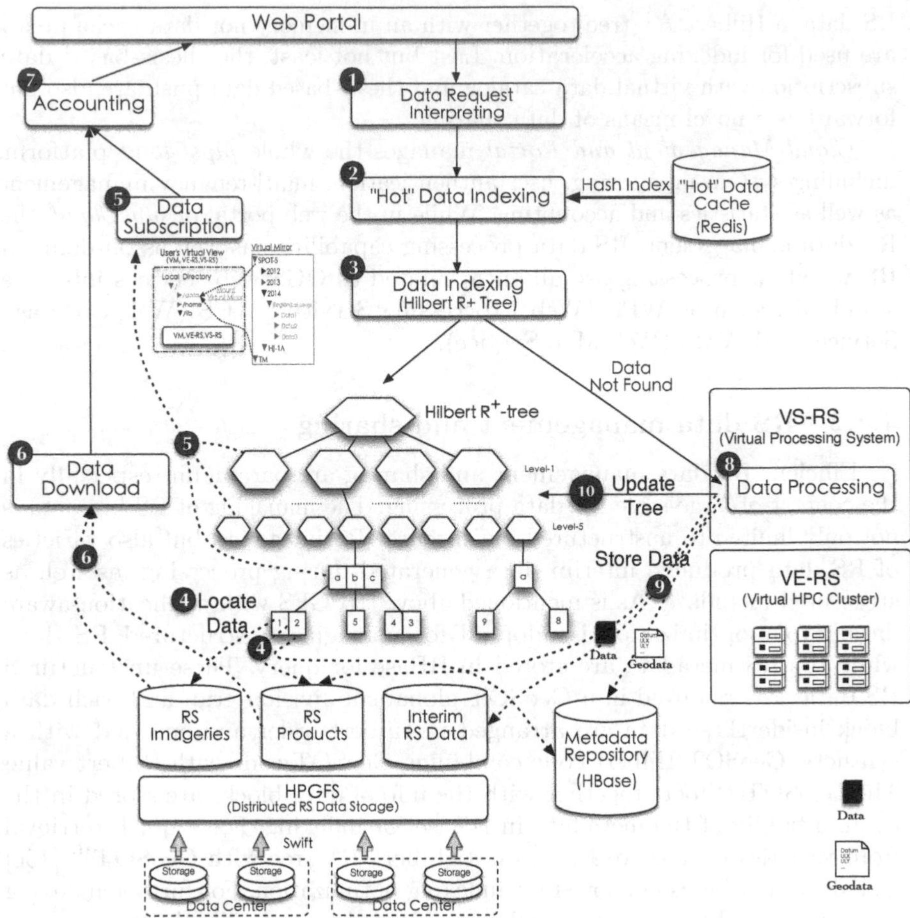

FIGURE 4.2: The runtime implementing of data management and sharing.

local mount point. Sixth, if an acquisition of data or metadata is needed, then it invokes GridFTP for downloading. Finally, Accounting is used for charging if the data is not free.

However, when the requested data products are not available then an RS data product processing could be requested to VS-RS. The interim RS data and final RS data products generated during processing would be stored to the interim and products repository based on HPGFS, while the relevant metadata will be abstracted and inserted into the metadata repository based on HBase. Meanwhile, the Hilbert R^+ tree should also be updated for further indexing, and the access RS data or metadata entry would automatically cache into Redis as "hot" data.

4.4.2.1 HPGFS: distributed RS data storage with application-aware data layouts and copies

The intensive irregular RS data access patterns of RS applications which always perform non-contiguous I/O across bunches of image data files would inevitably result in low I/O efficiency. Meanwhile, HPGFS which extends the prototype of OrangeFS offers an efficient and easy to use solution from the server side to natively support the direct distributed storing and concurrent accessing of massive RS data with different irregular I/O patterns. With the interfaces for data layout policy customization, distributed metadata management, OrangeFS is highly configurable. In HPGFS, a logical distributed RS data object model is adopted to organize the RS image datasets with complex data structure. It is worth mention that HPGFS offers I/O interfaces with RS data operation semantics together with application-aware data layout and copy policies associated with expected data access patterns of applications.

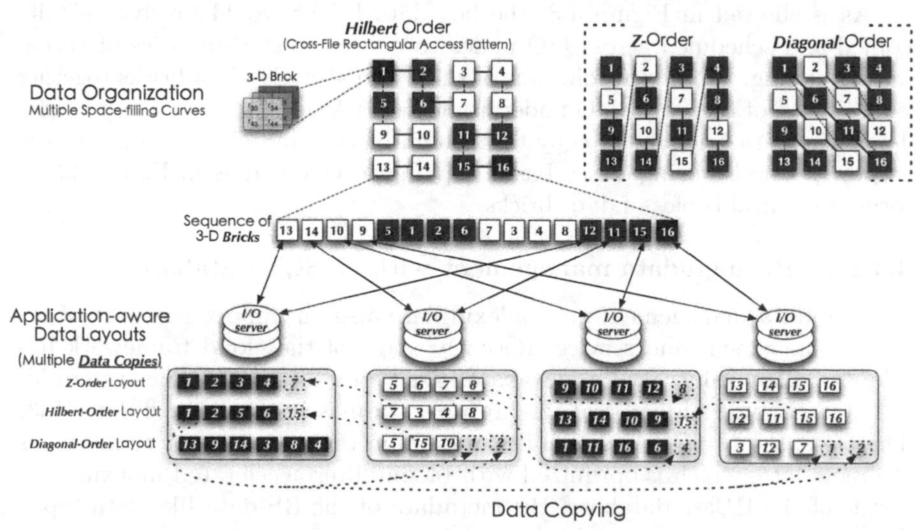

FIGURE 4.3: The application-aware data layout and data copies.

HPGFS adopts a logical distributed RS data object model to describe the multi-band RS image, geographical metadata as well as relevant basic RS data operations. The light-weighted Berkeley DB in distributed metadata servers is adopted for storing, managing and retrieval of the complex structured geographical metadata in a key-value fashion, while the multi-dimensional (normally 3-D) images are sliced and mapped into a 1-D data array using multiple space-filling curves simultaneously. Then the arrange data array is scattered over a number of I/O servers with application-aware data layout policies. The basic RS data operation includes some metadata inquiry interfaces and geographical operations like projection reform and resampling.

As is depicted in Figure 4.3, application-aware data layouts and data copy policies consistent with expected RS data access patterns are adopted for optimal data layout and exploiting data locality. It is worth noting that multiple redundant data copies with different application-aware data layouts are all simultaneously pre-created for each individual RS data. Instead of the data striping method with fixed or variable stripe size, RS data are sliced into data bricks, which are also multi-dimensional non-contiguous data. By awareness of the expected I/O patterns of RS applications, the 3-D data bricks in each copy of data are mapping and organized using a Space-filling Curve that best fits some certain data access patterns. Data copy organized in the Z-order curve is provided for the consecutive-lines/column access pattern, diagonal curve is for diagonal irregular data access pattern, while Hilbert curve is used for the rectangular-block access pattern. With the knowledge of the I/O patterns, the requested RS data region distributed across different datasets or even data centers can be organized and accessed locally in one single logical I/O.

As is showed in Figure 4.3, the hot data bricks would be dynamically copied and scheduled across I/O nodes to adhere to the statistics of actual data accessing. During the idle time of the I/O nodes, the data bricks together with the list of the target I/O nodes would be packaged as a "brick copy task". Then, the data brick in the copy task would be copied and transferred to the target I/O nodes using a tree-based copying mechanism as in Figure 4.3 to form dynamical copies of data bricks.

4.4.2.2 RS metadata management with NoSQL database

Metadata management and indexing are also an importance part of RS data management and service. Recently, most of the cloud framework has adopted a key-value NoSQL database based on a distributed file system for the storage and random, real-time retrieval of massive structured data [133]. Therefore, the NoSQL approach is employed in the management of enormous RS metadata, which is optimized with bloom filter accelerated indexing. By virtue of the HBase database, the metadata of the RS data like data types, data products, and geographical metadata are organized and stored in the column family. The thumbnails of RS data for quick view are also stored in HBase in the forms of several compressed tiles that are sliced with fixed size. The actual data of the HBase database are stored in an HDFS distributed file system in the Cloud. Following this approach, the geographically distributed metadata along with unstructured figures could also be managed by HBase for on-line metadata retrieval and quick view.

With the proliferation of data, the amount of metadata and catalogs would be bound to soar up. Not surprisingly, there would be millions of data entries in the metadata table of HBase. Actually, these enormous data entries in key-value fashion are normally serialized into thousands of files. Therefore, the on-line retrieval of metadata at extreme volume would definitely be a disaster. Actually, most of the key-value database are indexed by keyword of the data entry. There are several common indexing data structures, including Hash-based indexing, R-tree indexing, B-tree indexing and so on. The indexing trees are normally used for local indexing inside a key-value storage, while the Hash mechanism is used for locating the data nodes. However, in this scenario, a global indexing tree would not be a wise choice, since the cost of building and maintenance could be unprecedentedly huge.

In the distributed scenario of *pipsCloud*, a hybrid solution of combining both R-tree indexing and bloom filter is adopted instead of a global data indexing. Actually, we build a local R-tree indexing only for the metadata entries serialized in a group of data files or located in a single data center, instead of a global one. Meanwhile, a bloom filter is a space-efficient probabilistic data structure that is employed for global Hash-indexing; examples are Google BigTable and Apache Cassandra. Here, it is used to test whether the requested data belongs to a specified group of files or is located in a certain data center. In the case of metadata retrieval, the first step is to determine which group of entries it belongs to by conducting multiple bloom filter indexing of each group concurrently in parallel. Then follows a local search along an R-tree of the selected group of metadata entries. Eventually, the faster decision of the bloom filter as well as the k independent Hash lookups in parallel, would give rise to an accelerated metadata query through HBase and reduced costs of global indexing.

4.4.2.3 RS data index with Hilbert $R^{+}tree$

Quick and efficient data retrieval among enormous distributed RS data has always been a considerably challenging issue. Historically, indexing data structures like R-tree or B-tree are normally used to improve the speed of global RS data retrieval and sharing. Actually, the performance of the indexing tree greatly depends on the algorithms used to cluster the minimum bounding rectangles (MBRs) of the RS data on a node. Hilbert $R^{+}tree$ employs a Hilbert space-filling curve to arrange the data rectangles in a linear order and also group the neighboring rectangles together. To meet the requirements of real-time RS data updating, a dynamic Hilbert $R^{+}tree$ is normally more desirable. But compared to a static tree, it is also relatively more time-consuming and complicated.

Normally, RS data products are subdivided into standard scenes according to some global subdivision grid model. A data scene may span several degrees of longitude in lat-long geographical coordination, like 5° for Modis data. Under this consideration, *pipsCloud* adopts a hybrid solution, which the combines the Hilbert $R^{+}tree$ with a GeoSOT global subdivision grid model. As is showed in Figure 4.4, the global RS data are first grouped through a deferred quad-subdivision of GeoSOT global subdivision grid model. Normally, through several levels of quad-subdivision (like 5 levels) a quad-tree could be constructed, while if the a RS dataset covers a really big region that much larger than the GeoSOT grid, then this dataset would be logically further divided into

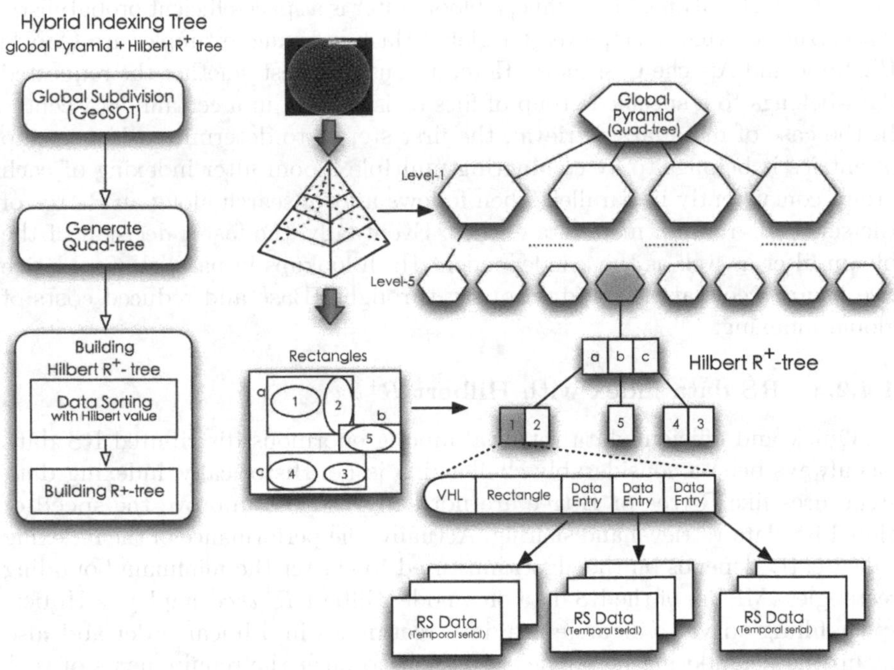

FIGURE 4.4: Optimal RS data indexing with Hilbert $R^{+}tree$ and global subdivision grid.

data blocks. Then the RS datasets or data blocks inside the geographical region of each leaf node of the quad-tree would be re-arranged according the Hilbert value of the center of the rectangles (i.e., MBR of the RS data or blocks). Following this method, each RS dataset or data block would be encoded with a unique GeoSOT-Hilbert code which consists of both the GeoSOT code and Hilbert value. Given the Hilbert ordering, we generate new tree nodes and assign rectangles of RS data to these tree nodes sequentially. The non-leaf node contains LHVs (Largest Hilbert Value) and also the geographical region of the rectangles. Then by recursively sorting these new nodes by the Hilbert value of its rectangle and creating new nodes with a higher level, a dynamic $R^{+}tree$ could be generated. The leaf node of this hybrid Hilbert $R^{+}tree$ is a data entry node, which contains the information (URL) of a temporal serial of RS data that is inside the rectangle of node.

With the unique GeoSOT-Hilbert code, the Hilbert $R^{+}tree$ and the RS metadata repository could be easily connected. For each RS dataset indexed in the leaf node of the Hilbert $R^{+}tree$, the GeoSOT-Hilbert code as well as MBR of the data itself and the data blocks inside this data would all be stored together with metadata in HBase. Accordingly, the searching of a given data region would be started from the root, descend the quad-tree of the GeoSOT global grid model, and then visit the nodes in the Hilbert $R^{+}tree$ that intersect the desired rectangle to get the GeoSOT-Hilbert code of the data so as to access metadata in HBase and acquire imageries with the URL.

4.4.2.4 RS data subscription and distribution

Data sharing is paramount especially in the scenario where enormous RS data are geographically scattered across data centers. *pipsCloud* provides on-demand data sharing to a wide range of users in a data subscription manner. Through data subscription, the required RS data retrieved by the search condition can be re-arranged and virtually mounted to the local storage of the user's virtual machine or virtual environment as local catalogs (Figure 4.5). Unlike conventional ways of data sharing, the subscribed RS data could be accessed and shared in a more easy-to-use and intuitive way through a virtual data catalog mounted locally. Once there is an up-to-date distribution of the subscripted data, data users may be informed and choose to update the virtual data catalog mirror without extra data retrieval or downloading.

The flow of RS data subscription is as follows: *Image Data Retrieval*: Users can get the list of the request RS data through data retrieval. Normally different ways of data retrieval are also offered, including inquiry condition and visible data selection through a map. The inquiry condition could be the synthesis of satellite, sensor, data product type, resolution, spatial region of data and time span. *Constructing Virtual Data Catalog Mirror*: Re-organize the retrieved RS data as in user or application customized catalog fashion, then link these data to form a virtual data catalog mirror. *Mount Virtual Data Catalog Mirror*: Take this catalog mirror as a shared network disk space and virtually mount

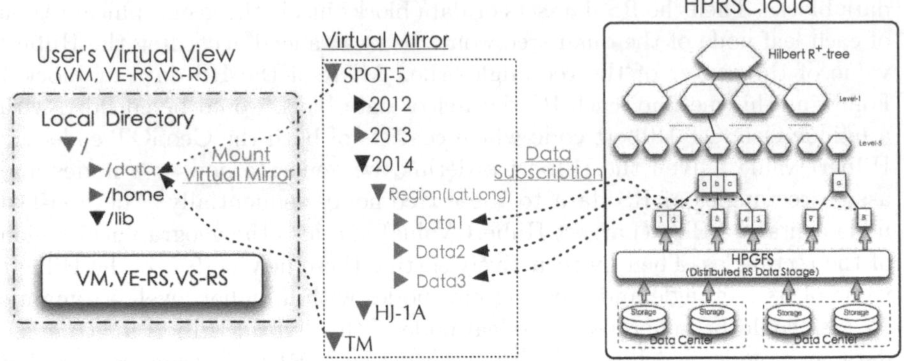

FIGURE 4.5: Data subscription and virtual catalog mirror.

it to the local disk of the user's VM or VE. Accordingly, data users could access any of the RS data inside the virtual data catalogs without extra data downloading operations.

Quick and efficient RS data distribution and browsing of geographically distributed massive multi-resolution RS image data is another challenging issue in a remote sensing cloud. *pipsCloud* offers the GeoSOT global subdivision grid (GSG) model together with the Hilbert Space-filling curve and code to create a Hilbert global image pyramid. Based on the global image pyramid, the MapServer[11] is employed as the major platform for on-line RS data distribution and map browsing. For performance efficiency, TileCache[12] is used for caching the requested hot image tiles. Finally, the image data are published as a WMS service for map displaying and retrieval by a web-based map client like OpenLayers[13] or Ka-Map[14]).

4.4.3 VE-RS: RS-specific HPC environment as a service

VE-RS offers an RS-specific cluster environment as a service on top of the OpenStack enabled cloud framework. Based on the VMs or BMs provided by the cloud framework, VE-RS could build a virtual HPC cluster through automatic deployment of cluster auto-configuration tools like AppScale. By means of SlatStack, VE-RS allows customized deployment of RS software such as ENVI, GDAL and ERDAS, together with HPC tools like MPI, MapReduce, Torque and Ganglia on VMs or PMs in the virtual cluster. Furthermore, *pipsCloud* also provides easy-to-use interfaces for auto-deployment of an RS-specific HPC cluster with the APIs of ApppScale and SlatStack.

[11] MapServer, http://www.mapserver.org
[12] TileCache, http://tilecache.org
[13] OpenLayers, http://en.wikipedia.org/wiki/OpenLayers
[14] Ka-Map, http://ka-map.maptools.org/index.phtml?page=home.html

For efficient managing of RS big data, VE-RS adopts a parallel file system named HPGFS especially for RS imageries. To solve the poor I/O performance introduced by intensive irregular I/O patterns, HPGFS adopts application aware data layout policy so as to exploit data locality and reduce data movements. Moreover, for easy but productive programming, VE-RS provides RS-GPPS, generic parallel skeletons for large-scale RS data processing applications on the MPI-enabled cluster environment. It also adopts a distributed RS data structure with fine-designed data layout control across distributed memories for efficient loading and communicating of RS big data. In addition, VE-RS adopts Torque scheduler as a local resource manager and scheduler in the virtual cluster, and ganglia for system monitoring.

When a virtual HPC cluster is requested, the generation and auto-deployment of VE-RS is invoked as showed in Figure 4.6. Firstly, the VE-RS customization request is parsed into a detailed resource requirement, such as the number or type of CPU, memory, network as well as cloud storage. Secondly, it retrieves available BMs or VMs in the BMs/VMs repository. Thirdly, if the existing BMs/VMs could fit the requirement then it creates it according to user needs, including physical machine scheduling by nova-scheduling, BM/VM image deploying, employing network by nova-network, mount cloud storage from Cinder or Swift and then registers it to the BMs/VMs repository. Fouthly, it employs cluster auto-configuration tools like AppScale to automatically build HPC cluster with HPC tools like MPI/MapReduce, Torque and Ganglia. Fifthly, it uses SlatStack to automatically deploy a programming model for RS, and other RS software like ENVI, GDAL and ERDAS. Following this process, a RS-specific cloud-enabled cluster environment would be automatically generated and registered into the VE-RS repository for later use.

4.4.3.1 On-demand HPC cluster platforms with bare-metal provisioning

In this study, we adopt xCAT to extend the dominant OpenStack platform for supporting bare-metal provisioning, and use the KVM hypervisor for VMs provisioning. OpenStack consists of a collection of software components, including Nova for computing resource (VMs/BMs) management, Glance for image management and Swift for building cloud storage. Normally, OpenStack basically offers virtual machines and the resource visualization is enabled by some hypervisors like Xen or KVM. As is shown in Figure 4.7, we use the KVM hypervisor for the creation and management of VMs from the pool of physical machines. When a virtual machine is requested, Nova-compute component of OpenStack will invoke the API of KVM for the creation and deployment of VMs, while, the virtualization and deployment of network resources is conducted by nova-network.

Actually, bare-metal provisioning is not directly supported in OpenStack. Hypervisor xCAT as a scalable distributed resource-provisioning tool, provides unified interfaces for discovery and software deployment of physical

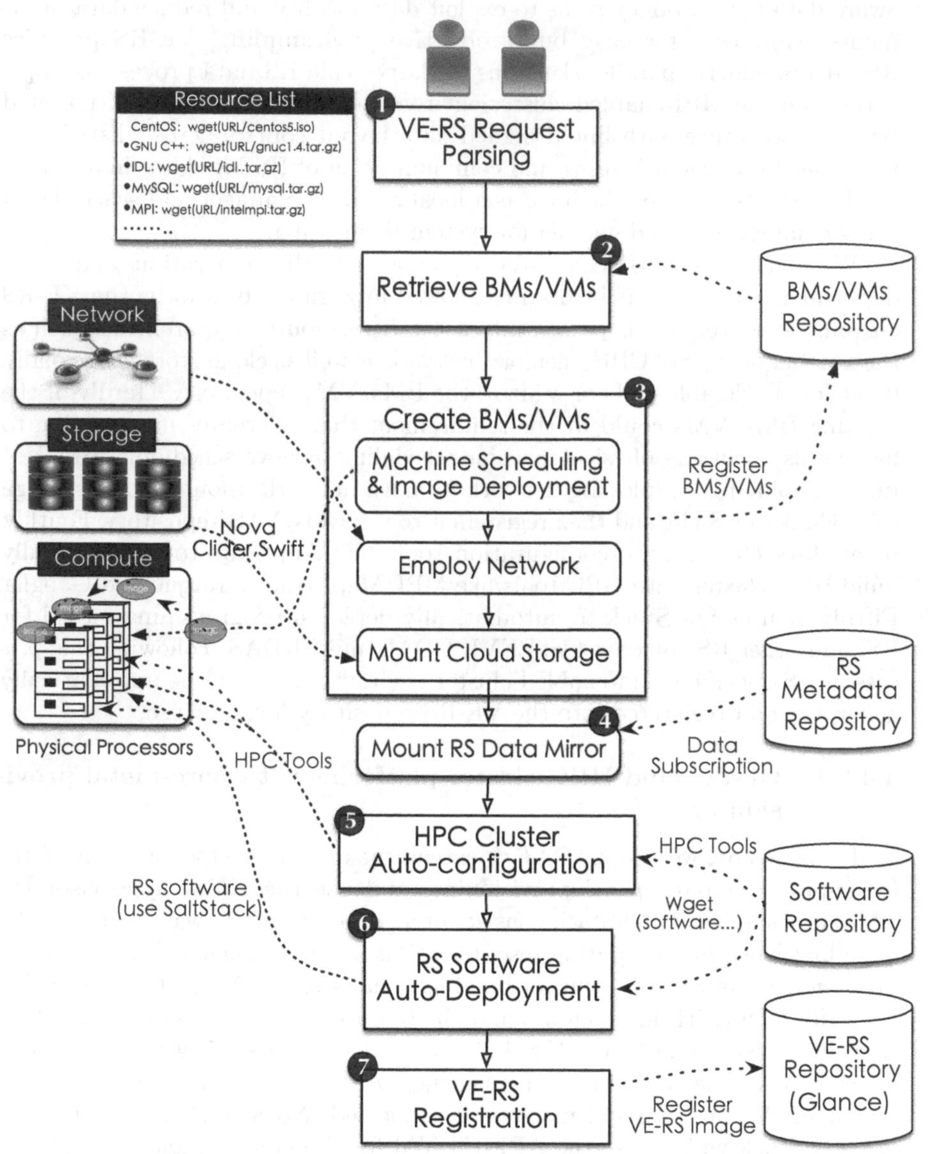

FIGURE 4.6: The generation and auto-deployment of VE-RS.

machines. However, to enable bare-metal machines in OpenStack through xCAT, a bare-metal driver for xCAT should be integrated in the Nova-compute component ([102]). Normally, Nova-compute uses the libvirt library to manage different hypervisors for diverse virtualization approaches. In this context, the bare-metal driver for xCAT is needed as an alternative to the libvirt driver. On one hand, the xCAT driver deals with the bare-metal (BM) machine requests from Nova-compute, and on the other hand it communicates with xCAT to complete resource provisioning.

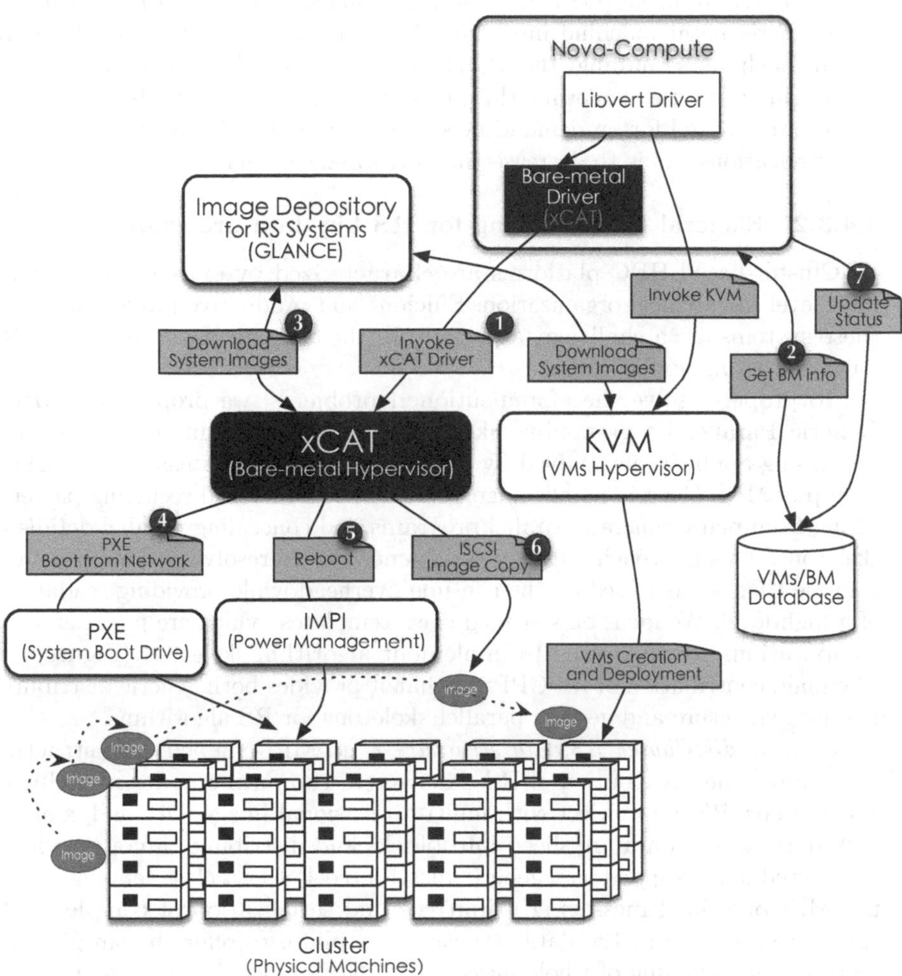

FIGURE 4.7: Bare-metal provisioning in Cloud with xCAT.

As is shown in Figure 4.7, when an HPC cluster environment for RS data processing with bare-metal machine is requested by a user, the nova scheduler will choose a nova-compute node and pass the request to the Libvert driver of it. Then, the Libvert driver would invoke the so implemented bare-metal driver of xCAT and transfer the request to xCAT. Consequently, the xCAT will take charge of everything. It gets the information of bare-metal machines from BM/VMs database, and downloads the system images with OS and software needed for building the HPC cluster for RS. Then, xCAT activates the boot loader of the physical machine using PXE (Preboot Execute Environment), and power on the machine with power management driver IMPI. After that, xCAT boots the physical machine from the network and deploys it with the specified system image (OS and software). Finally, it registers the information of the bare-metal machine into the BM/VMs database. In case the bare-metal machine is running, the xCAT is also responsible for managing and monitoring its status. Following this process, not only VMs but also BMs could be accommodated for on-demand needs of variable HPC cluster platform for RS applications, so as to decrease the performance penalty.

4.4.3.2 Skeletal programming for RS big data processing

Cluster-based HPC platforms are characterized by extreme scale and a multilevel hierarchical organization. Efficient and productive programming for these systems are a challenge, especially in the context of data-intensive RS data processing applications.

To properly solve the aforementioned problems, we propose RS-GPPS, Generic Parallel Programming Skeletons for massive remote sensing data processing applications enabled by a template class mechanism, and working on top of MPI. Generic parallel algorithms are abstract and recurring patterns lifting from many concrete parallel programs and concealing parallel details as skeletons. This approach relies on type genericity to resolve polymorphism at compile time, so as to reduce the runtime overhead while providing readability of a high-level. We focus on so-called class templates, which are parameterized computations patterns used to implement algorithm skeletons (Figure 4.8). The main contribution of RS-GPPS is that it provides both generic distributed RS data structure and generic parallel skeletons for RS algorithms.

Generic distributed RS data structure is an MPI-enabled data structure that allows the distribution of RS data across computing nodes in a cluster. The massive RS data object with multi-dimensional image data and complex metadata, whose data are sliced into blocks and distributed among nodes is abstracted and wrapped as a generic distributed RSData class template. Also the MPI one-sided messaging primitives and serialization of complex data structure are used in the data structure template, to offer the simple data accessing and residing of whole massive RS data in distributed memory space among nodes like local ones.

Generic parallel skeletons for RS algorithms that perform computations on distributed RS data structures could be used for RS algorithms with different computation modes. These skeletons express parameterization of parallelism without concern for implementation details like data distribution and task

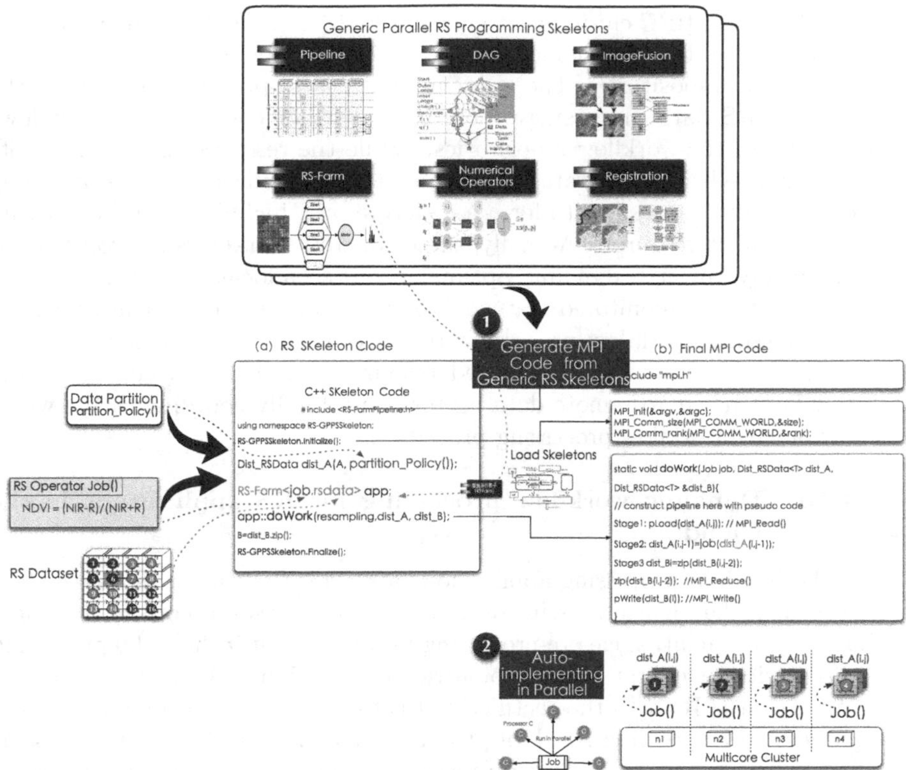

FIGURE 4.8: The skeletal parallel programming model for RS big data in pipsCloud.

partition, complicated access modes of RS data, and all low-level architecture dependent parallel behaviors. When a generic parallel skeleton is instantiated and declared, the computations on distributed remote sensing data objects are performed. Firstly, the task would be divided into subtasks by a two-stage task partition strategy, first nodes then intra-nodes, which are consistent with the data partition strategy. Then it actually loads the data blocks owned by each node concurrently through the parallel I/O operations enabled by parallel file system. Finally, the user defined remote sensing sequential code encapsulated in the job class would be implemented in parallel by each process. In this situation, the ease of parallel programming could be offered with a minimum concern for architecture-specific parallel implementation behaviors.

4.4.4 VS-RS: Cloud-enabled RS data processing system

VS-RS offers on-demand workflow customization and dynamic processing for various large-scale RS applications in the Cloud as on-line services on top of a cloud-enabled HPC cluster environment VE-RS. It consists of order manager, resource scheduler, runtime for collaborative workflow processing, and data or algorithm repositories. The order manager is responsible for parsing the requested RS data processing orders into abstract collaborative workflows according to the workflow repositories. While the resource scheduler adopts an optimal scheduling strategy to conduct an optimized resource mapping for the abstract workflow to form a concrete one, including data, algorithm and computing resources. Actually, these concrete workflows are constructed dynamically through dynamic optimal resource allocation during runtime according to the monitored status of resources and system. Meanwhile, the Kepler-enabled workflow processing runtime dynamically implements each step of the workflow with allocated resources on the local cluster in VE-RS or launches it to remote data centers, and finally coordinates the whole collaborative workflow processing procedure.

4.4.4.1 Dynamic workflow processing for RS applications in the Cloud

The RS data processing applications are typically somewhat on-the-flow processing. The whole processing procedure consists of several processing stages. Take a typical multi-stage pre-processing for instance; it includes L0 processing, radiometric correction (RC), geometric rectification (GR) followed by fine rectification (FR) or ortho-rectification (OR). Each of the processing steps produces corresponding RS data products. This kind of RS data processing workflow generates a data-driven processing flow, each step depending on the output data of the preceding step as input data.

In a traditional RS data processing system, the workflows for various RS applications are always predefined as static ones. But many RS applications normally demand on-demand workflow processing capability. In this scenario, the dynamic customization of application-specific workflows according to the variable needs is essential. Moreover, the RS observation data generally are scattered among different satellite data centers geographically. So large-scale RS applications like regional to global drought monitoring ([134]), normally need a collaboration of several data centers. In this sense, not only the data, but also the processing workflows from other data centers or scientists need to be shared and cooperated. To complicate the situation the unstable computing environment across data centers will inevitably lead to the failure of whole processing. Therefore, dynamic resource allocation and scheduling are essential for distributed workflow collaboration across data centers.

To solve the dynamic RS workflow processing issue, we put forward a two-level workflow processing scheme with both the abstract workflow and the concrete one. The abstract workflow is used to logically represent the complex processing procedure customized by domain scientists. Each processing step in the abstract workflow is a logical function rather than an actual processing program, whose actual algorithm and data resources are not decided yet. When an abstract workflow is customized, it would be expressed in XML format and stored in the RS workflow depository, while, the concrete workflow is not built at once but dynamically constructed and implemented by an efficient workflow engine through dynamic resource mapping during runtime. In the case when one step of the abstract workflow is allocated with the required algorithm, data and computing resources then it will be launched to designated nodes or a remote data center for parallel implementing and collaboration by the workflow engine. Following this process, the large-scale RS application could be dynamically implemented on the HPC cluster or across data centers for global workflow collaboration with optimal runtime resource allocation according to resource status.

For dynamic workflow processing with high efficiency, a proper workflow engine is also of vital importance. Currently, scientific workflows ([135]) are gradually employed to formalize and enable distributed scientific processing in various disciplines, such as physics and earth science. Compared to the traditional control flow oriented workflow system, scientific workflow management systems (SWFMSs) like VIEW ([135], Kepler ([136] and Pegasus ([137]) are typically data flow oriented. In this study, we adopt the Kepler engine that is most widely used as the main scientific workflow runtime and management system.

On-demand workflow customization serviced by VS-RS in *pipsCloud* is demonstrated in Figure 4.9. When an RS processing workflow creation or customization is requested through a simple drag-and-drop of algorithms in a graphical user portal, a searching operation is triggered in the RS algorithm repository for a name catalog of various registered RS algorithms. Then follows the composition of user selected algorithms together with the control logics among these algorithms to form an abstract workflow. Here the algorithms which form workflows only refer to the functional names rather than the real algorithm resources with the executable program. The customized abstract workflows expressed in XML format are then registered into the RS abstract workflow repository for further processing.

Dynamic RS workflow runtime enabled by Kepler workflow engine and a two-level workflow scheme is illustrated in Figure 4.9 for collaborative large-scale RS workflow processing across data centers or clouds. In the case when a RS data products processing order is requested by a user through a portal, then an RS data processing procedure would be launched on the RS workflow runtime for processing.

The dynamic processing of large-scale collaborative workflows on Kepler-enabled runtime goes as follows:

- Firstly, *Abstract workflow matching* is responsible for interpreting the requested orders into abstract RS workflows without allocation. With the key word "Product Type", runtime searches for the corresponding abstract workflow in the workflow repository for each RS order requested through the cloud web portal, while, the abstract workflow interpreted only tells the blueprint of the data processing procedure, including the

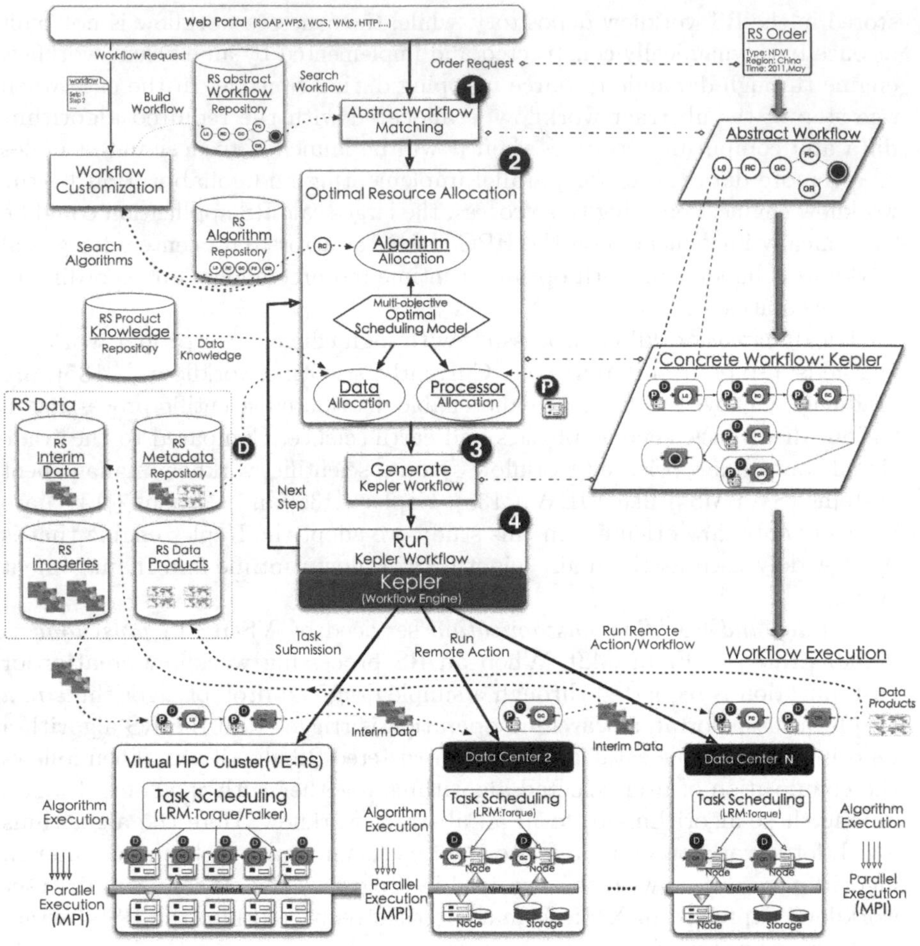

FIGURE 4.9: Dynamic and optimal workflow processing for large-scale RS applications with Kepler.

functional name of each step as well as the control logic among them. But the actual processing program or data needed for processing in each workflow step is not decided yet.

- Secondly, *Optimal Resource Allocation* continues to conduct optimal resource mapping for each workflow step according to the current status of various resources and systems. The resources here refer to three main categories of resources including algorithm resources with actual programs, various RS data required for processing and also processing resources like processors and network which are needed for execution. Initially, a knowledge query from product knowledge repository is invoked for acquiring the knowledge rules for this designated RS data product. The product knowledge represents in rules that mainly indicate the requirement of the RS data, such as the resolution or sensor of the RS imageries, auxiliary data as well as some parameter data or file needed. Then with the knowledge rules of data products, there follows the generating of a condition statement for an RS data query. Accordingly, a list of RS imageries or auxiliary data could be drawn out from the data repository for further processing. After that goes the algorithm allocation, the candidate executable programs are deposited from the algorithm repository with the key word of the functional name of the algorithms. In addition, the available computing resources like processor and network are also allocated for processing according to the current status through monitoring.

However, the problem worth noticing is how to choose the resources that fit the data processing procedure best so as to achieve an optimal performance QoS target. The main reason for that would be the abundance of candidate programs of algorithms, RS data as well as computing resources meeting the requirement of these specified processing steps in the workflow of data products processing. Nevertheless the location of RS data, network bandwidth and usage as well as the CPU capacity and usage of processors are all factors that affect the total performance QoS target of the whole workflow.

To achieve optimal resources for a certain step of workflow, the workflow runtime employs an optimal resource scheduling model with performance QoS target function and user-customized scheduling rules. The QoS target function is a target function which takes the data amount, network width, program performance and processor performance into consideration, each factor of which is assigned with an empirical weight value. Moreover, the scheduling rule could also be customized and stored into a rule repository in a key-condition-threshold style. Some basic scheduling rules like near-data computing for reducing data movement are also provided in the workflow scheduling for optimization. According to the performance QoS target function and scheduling rules, we could select the best-fit RS data, algorithm programs and computing resources from candidates with an optimal final performance QoS. Then these resources are allocated to this certain processing step of the workflow.

- Thirdly, *Partly generating concrete Kepler workflow* from abstract workflow with allocated resources. Here runtime only generates part of the Kepler workflow for certain processing steps with allocated resources. Each step of the Kepler workflow is then represented as an executable Kepler "actor".
- Fourthly, *Run Kepler workflow* on Kepler workflow engine. In the case when the processing step of workflow is a local actor, then a PBS/Torque

task submission is triggered to the LRMs (Local Resource Manager). Then the LRMs launches the program of this workflow processing step onto the allocated processors (VMs or BMs) in the virtual HPC cluster environment in the Cloud and executes it in parallel, while, if the workflow step is "sub-workflow" expressed as a web service actor, Kepler would directly invokes the web service interfaces for execution. If the processing step is a remote job execution, then a remote action is invoked with a remote job submission. After receiving the job submission, the LRMs of the remote data center would soon run the program on processors and final feedback with interim data. The interim data would be cached into the interim data repository so as to prepare for data transferring to the next data center or cloud for further processing of workflow. As is shown in Figure 4.9, in the processing workflow of producing global NDVI data products, runtime firstly executes two initial steps of L0 processing and geometric correction (GC) on the virtual HPC cluster of cloud, and then passes the interim data products to Data Center 2 and launches a remote execution of radiometric correction (RC), when the RC is finished; then Kepler triggers a remote job submission to the LRMs of data center N for parallel implementing of the last two programs.

- Finally, the workflow processing is continued recursively from optimal resource allocation, generating Kepler workflow to implementing workflow collaboratively until the end of the workflow procedure. Following this process, the entire complex processing workflow could be generated and implemented dynamically on the Kepler engine with a nearly optimal performance QoS of the whole processing procedure. When the workflow ends, the RS data products would be registered into the RS data product repository for downloading.

Consequently, with the logical control and data transferring among data centers, a distributed workflow among different data centers or cloud systems could be collaboratively implemented. Each step of the workflow is implemented with the optimal allocated resources according to current system status. Even when a failure of the allocated resources occurs, then a re-allocation of the resource would be triggered for a re-build and re-run of the Kepler workflow.

4.5 Experiments and Discussion

The *pipsCloud* which offers a high-performance cloud environment for RS big data has been successfully adopted to build the Multi-data-center Collaborative Process System (MDCPS). By virtue of the data management service in pipsCloud, the multi-source raw RS data, interim data and also data products can all be efficiently managed and accessed. Through the VE-RS

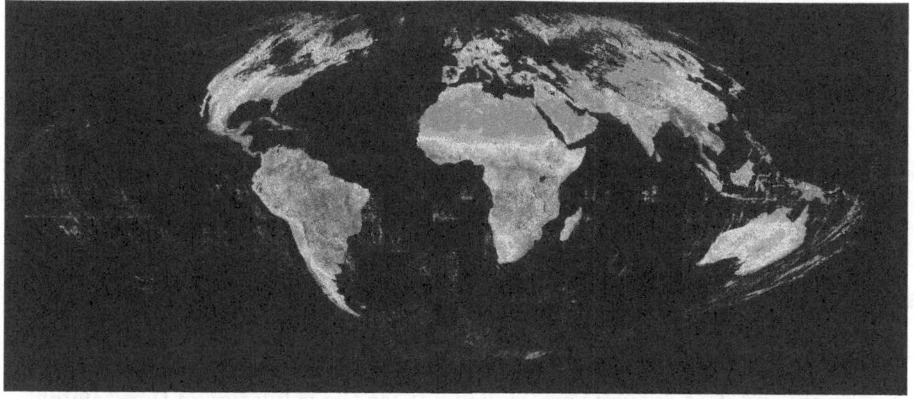

FIGURE 4.10: The 5-day synthetic global NDVI products in 2014.

(a) NPP products from day 211 to 215

(b) NPP products from day 221 to 225

FIGURE 4.11: The 5-day synthetic Global NPP products in 2014.

service in pipsCloud, a customized virtual HPC cluster environment is easily built and equipped with RS software, a parallel programming model and a large-scale task scheduling especially for RS applications. By employing the VS-RS service offered in pipsCloud, MDCP are well constructed and equipped upon the VE-RS cluster environment with order management, a workflow engine and a depository for RS algorithms and workflows. Furthermore, enabled by the Kepler workflow engine, the complex processing procedures for global RS data products are customized as dynamic workflows that are implemented through collaboration a cross multiple data centers. This processing is dynamic since the concrete workflows are not predefined but dynamically formed through runtime resource mapping from abstract workflows to data centers.

Actually, MDCPS connects several national satellite data centers in China, such as CCRSD[15], NSOAPS[16], NMSC[17]. It offers on-line processing of regional to global climate change related quantitative RS data products with these multi-source RS data across data centers. The RS data products generated by MDCPS include vegetation related parameters like NDVI [18] and NPP[19], radiation and hydrothermal flux related parameters like AOD[20] and SM[21], as well as global ice change and mineral related parameters. The 5-day global synthetic NDVI parameter product in 2014 generated using MODIS 1km data is shown in Figure 4.10. The 5-day global synthetic NPP parameter products which were also produced with MODIS 1km data in day 211 to 215 and day 221 to 225 in 2014 are relatively demonstrated in sub figure (a) and (b) in Figure 4.11.

The performance experiments on typical RS algorithms with both increasing processors and data amounts are carried out for the validation of the scalability of the pipsCloud platform. In this experiment, two MPI-enabled RS algorithms are chosen for implementing, including NDVI and NPP. Meanwhile, the pipsCloud platform offers a virtual multi-core cluster with 10 nodes connected by a 20 gigabyte Infiniband network using RDMA(Remote Direct Memory Access) protocol. Each node is a bare-metal provisioned processor with dual Intel (R) Quad core CPU (3.0 GHz) and 8 GB memory. The operating system was Cent OS5.0, the C++ compiler was a GNU C/C++ Compiler with optimizing level O3, and the MPI implementation was MPICH.

The runtime and speedup performance merit of both NPP and NDVI with increasing numbers of processors are illustrated relatively in Figure 4.12 and Figure 4.13. As is demonstrated in sub figure (a), the run time merit curves of these two algorithms decrease almost linearly especially when scaled to less than 4 processors (32 cores). However, the decrease rate is much slower when scaled from 5 processors (40 cores) to 10 processors (80 cores). The main reason for that would be the total run time which is relatively small makes the speedup not that obvious, since the system overhead could not be omitted. The same trend is also shown in sub figure (b) that the speedup metric curves of both two algorithms soar up linearly when scaling to 10 processors (80 cores).

With the amount of RS data increasing from 0.5 gigabytes to about 300 gigabytes, the experimental result is depicted in Figure 4.14. Judging from the performance curves demonstrated, the MPI-enabled NPP and NDVI algorithms implemented on pipsCloud both show their excellent scalability in terms of data.

---

[15]CCRSD: China Centre for Resources Satellite Data
[16]NSOAPS: National Satellite Ocean Application Service
[17]NSMC: National Satellite Meteorological Centre
[18]NDVI: Normalized Differential Vegetation Index
[19]NPP: Net Primary Productivity
[20]AOD: Aerosol optical depth
[21]SM: Soil Moisture

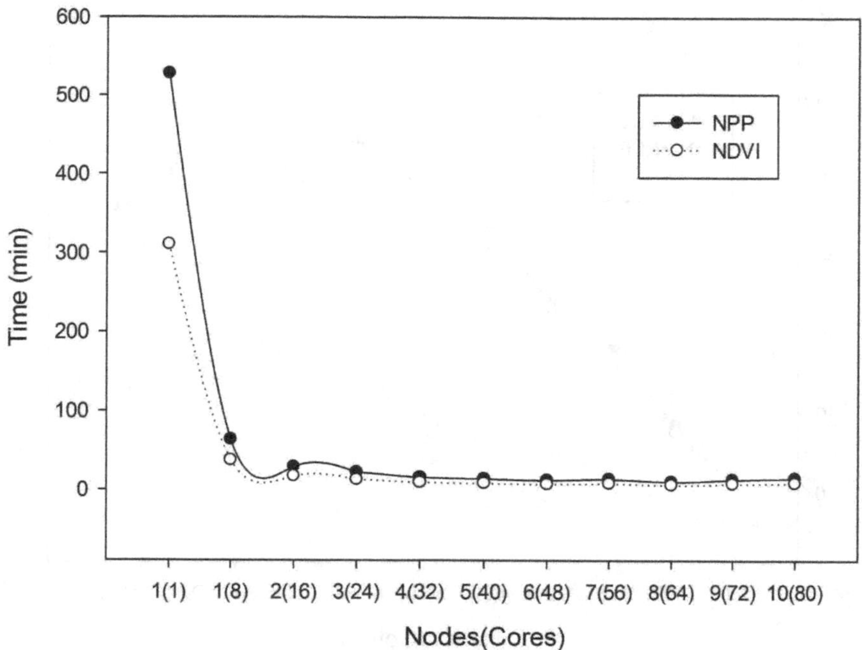

FIGURE 4.12: Run time of NPP and NDVI with scaling nodes.

4.6 Conclusions

The Cloud computing paradigm has been widely accepted in the IT industry with highly matured Cloud computing middleware technologies, business models, and well-cultivated ecosystems. Remote sensing is a typical information associated zone, where data management and processing play a key role. The advent of the high resolution earth observation era gave birth to the explosive growth of remote sensing (RS) data. The proliferation of data also gave rise to the increasing complexity of RS data, like the diversity and higher dimensionality characteristic of the data. RS data are regarded as RS "Big Data".

In this chapter we discussed how to bring the cloud computing methodologies into the remote sensing domain. We focus the overall high performance Cloud computing concepts, technologies and software systems to solve the problems of TS big data. *pipsCloud,* a prototype software system for high performance Cloud computing for RS is proposed and discussed with in-deep

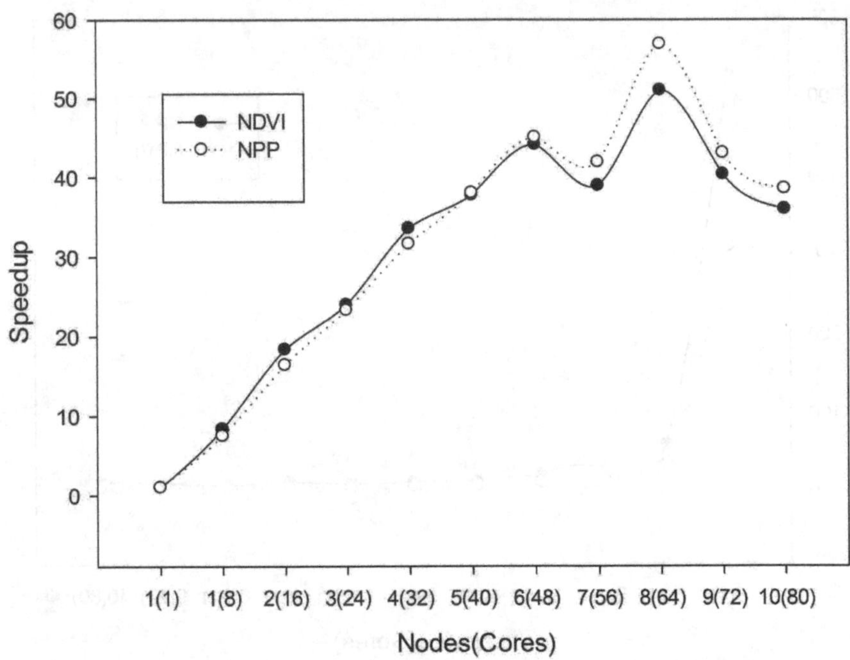

FIGURE 4.13: Speedup of NPP and NDVI with scaling nodes.

discussion of technology and implementation. As a contribution, this study brings a complete reference design and implementation of high performance Cloud computing for remote sensing.

In future applications, such as smart cities and disaster management, the great challenges will arise due to fusion of huge remote sensing data with other IoT data. The *pipsCloud*, benefiting from the ubiquity, elasticity and high-level of transparency of the cloud computing model, could manage and process the massive data, meeting the future applications' requirements.

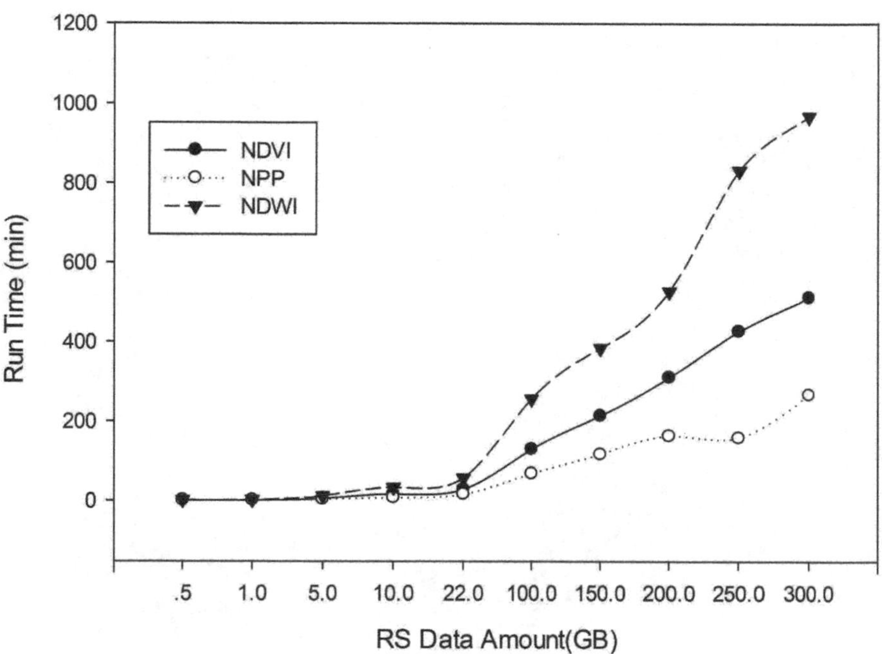

FIGURE 4.14: Run time of NPP and NDVI with increasing data amount.

Chapter 5

Programming Technologies for High Performance Remote Sensing Data Processing in a Cloud Computing Environment

5.1 Introduction

Recently, the amount of remote sensing data continuously acquired by satellite and airborne sensors has been dramatically increasing by terabytes per day, and the sheer volume of a single dataset is several gigabytes. Extremely massive data need to be processed and analyzed daily. The remote sensing datasets characterized by multi-band image data structure and abundant geographical information used for image processing are usually organized in various formats, making it rather trivial and difficult for algorithms to load and reside remote sensing data. With the proliferation of data, remote sensing data processing becomes extremely challenging because of the massive image data, the pixels of higher dimensionality, and various algorithms with higher complexity; also more complicated data access mode results from different dependences between computation and data. In particular, many time-critical applications like disaster monitoring even require real-time or near real-time processing capabilities.

The enormous computational requirements introduced by the unprecedented massive data and various complicated algorithms, especially many time-critical applications, have far outpaced the capability of single computer. Incorporation of multi-core cluster based high-performance computing (HPC) models in remote sensing image processing is an effective solution to address these computational challenges. However, the parallel cluster systems will be characterized by increasing scale and multilevel parallel hierarchy. To write efficient codes for parallel remote sensing algorithms usually involves handling data slicing and distribution, task partition, a message passing model among nodes and Shared memory model for cores, synchronization and communication with low-level APIs like the message passing interface (MPI). And it still remain a big challenge for a parallel system to process massive remote sensing data, as to load the massive datasets into memory at once before processing is no longer possible, and it is also quite complicated and inefficient to reside and communicate among nodes the datasets with multi-band images and complex structured geographical information. Therefore, Parallel programming on such systems for data-intensive applications like remote sensing is bound to be trivial, difficult and error-prone.

Generic parallel algorithms are abstract and recurring patterns lifting from many concrete parallel programs and conceal parallel details as skeletons. This approach relies on type genericity to resolve polymorphism at compile time, so as to reduce the runtime overhead while providing high-level readability. To properly solve the aforementioned problems, we propose RS-GPPS, Generic Parallel Programming Skeletons for massive remote sensing data processing applications enabled by a template class mechanism and work on top of MPI. We focus on so-called class templates, which are parameterized computation patterns used to implement algorithm skeletons. The main contribution of RS-GPPS is that it provides both a template for distributed RS data and skeletal parallel templates for RS algorithms. The massive RS data object with multi-dimensional image data and complex metadata, whose data are sliced into blocks and distributed among nodes is abstracted and wrapped as a generic distributed RSData class template. Also the MPI one-sided messaging primitives and serialization of complex data structure will be used in a template, to offer the simple data accessing and residing of whole massive RS data in distributed memory space among nodes like local ones. And the skeletal parallel templates are generic parallel algorithms performing computations on distributed RS data. These templates express parameterization of parallelism without concern for implementation details like data distribution and task partition, complicated access modes of RS data, and all low-level architecture dependent parallel behaviors. Only by writing user-defined sequential code which is also the type parameter of templates, and instantiating a skeletal parallel template, the RS parallel algorithms could be programmed as sequential ones with high parallel efficiency and a minimal runtime overhead of polymorphism.

The rest of this chapter is organized as follows. The next section reviews the related work, and the problem definition of massive remote sensing image processing is discussed in Section 5.3. Section 5.4 presents the design and implementations of our proposal of generic parallel skeletons for massive RS algorithms on MPI-enabled multi-core clusters. Section 5.5 discusses the experimental analysis of program performance, and Section 5.6 concludes this chapter.

5.2 Related Work

Clusters of multicore SMP (Symmetric Multi-Processors) nodes have been widely accepted in high performance computing and a number of programming paradigms are developed for this hierarchical architecture. The OpenMP[138] model is used for parallelization of shared-memory and distributed shared-memory clusters[139]. The Message Passing model (MPI)[140] is employed within and across the nodes of clusters. The hybrid programming paradigm MPI+OpenMP[141] exploits multiple levels of parallelism: the OpenMP model is used for parallelization inside the node and MPI is for message passing among nodes. However, developing parallel programs for hierarchical cluster architectures with low-level programming models and APIs, for example MPI and OpenMP, is still difficult and error-prone.

The skeletal parallel programming with higher-level patterns is adopted by researchers to simplify parallel programming. The eSkel[142] library provides parallel skeletons for C language. It can generate efficient parallel programs but leave users to handle low-level API with many MPI-specific implementation details. Lithium[143] is a platform-independent Java library and provides a set of predefined skeleton classes. Muesli[144] offers polymorphic C++ skeletons with a high level of abstraction and simple APIs. The programs can be produced by construction of abstract skeleton classes and deal with data transmission via a distributed container. However, it suffers a large overhead paid for runtime polymorphic virtual function calls. Google's MapReduce model[145], which supports Map and Reduce operations for distributed data processing in a cluster, is a simple yet successful example of parallel skeletons.

The generic programming approach uses templates to program generically. This concept has been exploited efficiently in the Standard Template Library (STL)[146], which has been extensively applied due to its convenient generic features and efficient implementations. Furthermore, the polymorphism is resolved at compile time because of its usual type genericity. The QUAFF[147] skeleton-based library offers generic algorithms. It relies on C++ templates to resolve polymorphism by means of type definitions processed at compile time. It reduces the runtime overhead of polymorphism to the strict minimum while keeping a high-level of expressivity and readability.

Low-level parallel programming models like MPI, OpenMP and MPI+OpenMP are extensively employed for remote sensing image processing. Examples include data preprocessing[148], mosaic[149], disaster monitoring[150, 151] and global changes. The aforementioned projects provide high-level pattern for parallel programming. However, these implementations did not develop their research for the massive remote sensing datasets with multi-band image data and complex structured metadata. Moreover, they did not handle the different dependences between computation and data of RS algorithms. In this situation, it remains a big challenge to program effective massive remote sensing data processing algorithms productively.

The RS-GPPS skeletons proposed in this study aim at addressing the above issues. They rely on the generic approach as QUAFF does to provide generic parallel algorithm skeletons for massive remote sensing processing. The RS-GPPS also develops its implementation for 1) massive RS data with complex data structure and 2) dependences between computation and data.

5.3 Problem Definition

This section presents the main issues related to parallel programming for massive remote sensing image processing applications on an MPI-enabled multi-core cluster. There are three aspects of this problem: massive remote sensing data, difficulties of parallel programming and data processing speed. The remote sensing image processing applications are overwhelmed with tons of remote sensing images. The first issue is related to the extremely large scale of remote sensing image data which can no longer possibly be loaded into memory at once, and how these data with multi-dimensionality and complex metadata structure can be easily sliced and communicated across nodes of cluster. The second issue is related to the programmability of parallel remote sensing algorithms with the great effort concerning parallel implementation details and the different dependencies between the computation and remote sensing data. The third issue is about data processing speed, namely how to sufficiently benefit from the multilevel hierarchical parallel architecture of multi-core SMP cluster, data I/O and data locality.

5.3.1 Massive RS data

The sheer size of remote sensing image datasets would be several gigabytes. And applications like global change and disaster monitoring are more likely to process big regional and even global multi-temporal, multi-band remote sensing image data from multi-sensors. Massive multi-dimensional remote sensing image datasets with large scale make the traditional way of processing inapplicable on clusters. It is no longer desirable to load the entire large datasets into the limited memory of a so called master node of a cluster, and distribute the sliced image data blocks among computing nodes before processing.

With the rapid development of remote sensing technology, the remotely sensed images from multi-spectral even hyper-spectral sensors usually have hundreds of spectral bands, such as the WIS sensor which produces 812 bands and the HJ-1A hyper-spectral imager 128 bands. The multi-band remote sensing images would always lead to the pixel's multi-dimensionality. On other hand, the RS datasets consist of abundant metadata including complex geographical information which is always involved in the data processing procedure of algorithms. Accordingly, to define and reside these RS datasets with multi-dimensional images and complex metadata structure in memory would be rather trivial. And the complex data structure of RS data also complicates the RS data slicing and communicating across nodes of cluster. Firstly, when the remote sensing image datasets are split into blocks, the geographical information of these blocks should be recalculated. For example, the latitude and longitude of the four corners of a data block in a geographic coordinate system should be recalculated using projection parameters, the geometric position of the data block, and the region of the entire image; this is rather cumbersome. In most cases, RS data will have to be split into partially overlapping blocks, as most algorithms do have to consider the dependence between computations and data, such as that the computation of each pixel depends on the neighborhood of that pixel. Secondly, the communication of the complex user-defined data types like RS image datasets across nodes is not properly supported by MPI. Programmers have to transmit the complex metadata information item by item via calling low-level MPI send receive communication API several times, which impacts negatively on parallel programmability.

5.3.2 Parallel programmability

With the extensive application of remote sensing images, there emerges a variety of algorithms with higher complexity. The dependencies between the computation and data vary with different algorithms, like data independent computation for pixel-based processing algorithms, region dependent computation for neighborhood-based processing algorithms and global dependent computation for global or irregular processing algorithms. These dependences will probably lead to different data parallelism, data computation modes, data/task partition strategies, and even complicated data access modes. Of course, these will make the parallel programming of RS algorithms more difficult. To code efficient parallel algorithms, programmers need to be proficient in the multilevel hierarchical parallel architecture of the multi-core cluster, and in dealing with both the message passing model for nodes and the shared memory model for intra-node. Moreover, programmers have to be tremendously concerned for lots of architecture-specific implementation details, like data communication and process synchronization with low-level message passing API like MPI.

Fortunately, many algorithms express parallelism with recurring parallel patterns. If the recurring patterns of parallel computation and communication could be abstracted and pre-implemented, these patterns could be reused by a collection of similar algorithms. In this situation, the ease of parallel programming could be offered with a minimum concern for parallel implementation behaviors.

5.3.3 Data processing speed

Processing of massive remote sensing data is extremely time-consuming. Especially the time-critical applications like Flood monitoring which requires being processed in real-time poses a big challenge to data processing speed. Therefore, the parallelization of remote sensing algorithms should concern how to take the parallelism of multi-core into account and sufficiently benefit from the multilevel hierarchical parallel architecture of the multi-core SMP cluster. And there remains a big gap between the performance of CPU and I/O. To complicate the situation is that the loading of massive RS image data before processing make the I/O consumption intolerable, eventually to become the performance bottleneck. So to greatly improve I/O performance and fully overlap the I/O operation with computation would also be critically important. In addition, if each node doesn't get all the data needed for processing, data exchanging would be incurred. The frequent data communication among nodes would also result in poor performance. For performance and energy efficiency it is critically important to take data locality into account.

5.4 Design and Implementation

We designed and implemented the generic parallel skeletons for massive remote sensing image data processing problems, RS-GPPS, to provide a more efficient and easy way to program remote sensing algorithms dealing with massive RS data on an MPI-enabled multi-core cluster. RS-GPPS consists of several generic parallel algorithm skeletons for remote sensing image data processing applications and the generic RS data type for distributed remote sensing image data objects.

Both the generic parallel algorithm skeletons and distributed RS data type are enabled by a template class mechanism. The data type of RS image data and user-defined data partition strategy function are exposed as parameters of the RS data type template. By specifying these parameters of templates, a distributed RS data object could be defined. Once the distributed RS data is declared, it would be sliced into blocks with a user-defined data partition strategy which could also be defaulted and distributed among nodes. The type parameters and member functions of these generic class templates are exposed to programmers as the interface of skeletons. As shown in Figure 5.1, only by specifying the user-defined job class which encapsulates the sequential code segments or functions as the parameter of the skeleton templates, could these generic parallel algorithm templates then be instantiated. And when these are implemented, every process locally runs the sequential function by calling user-defined job class. And if necessary the communications among nodes will also be implemented. Consequently, the low-level parallel programming details are shielded from users. The parallel remote sensing programs could be easily written as if they were sequential ones.

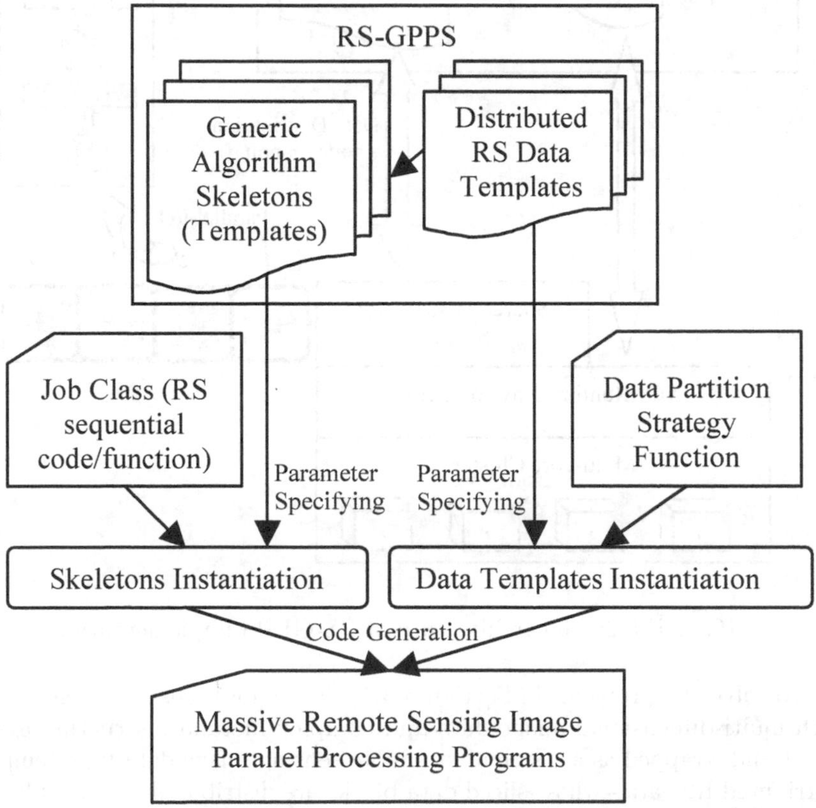

FIGURE 5.1: The flow of generating parallel programs for remote sensing data processing algorithms with RS-GPPS skeletons.

The system architecture of the RS-GPPS implementation is illustrated in Figure 5.2. The architecture-specific parallel details and the data slicing/distribution operations are pre-implemented in generic algorithm skeletons and distributed RS data templates respectively with MPI.

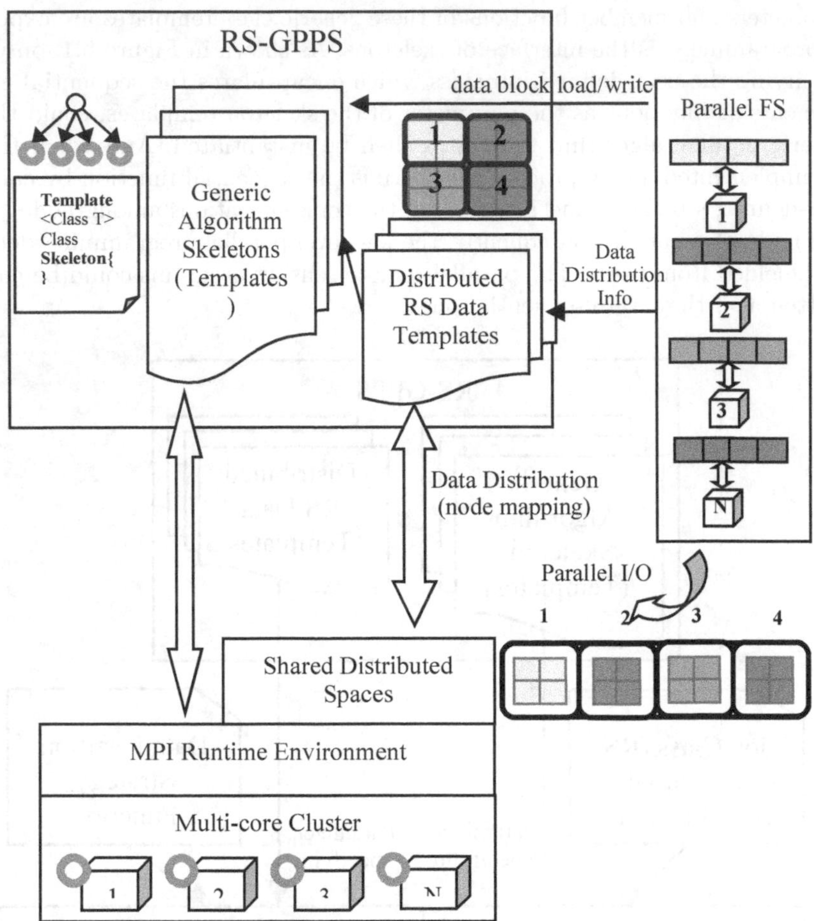

FIGURE 5.2: The architecture of RS-GPPS implementation.

To solve the problems in Section 5.3.1, the remote sensing image dataset with multi-dimensional image data and complex metadata structure are abstract and wrapped as a distributed remote sensing image data type template-distributed RSData - whose sliced data blocks are distributed across multi-core nodes of cluster. When a distributed RS data is declared, the entire RS dataset would be logically sliced into blocks and then assigned to nodes. The data slicing operation recursively splits the entire RS dataset into partially overlapped blocks with user defined or default data slicing strategies, and recalculates the geographical information. And then the sliced data blocks are assigned and distributed to nodes according to the map relationship of physical data blocks and I/O nodes in the parallel file system. Thereafter, through complex data type serialization, the sliced data blocks could be resided in the local memory of each node. And these memory spaces are exposed to all nodes for sharing and virtually converged to a global distributed RS data based on a runtime system with one-sided messaging primitives provided by MPI. Without any MPI two-side communication operations, the programmer could easily access any blocks of the entire distributed RS data object enabled by remote memory access primitives provided by MPI as if they are local ones.

To solve the problems discussed in Section 5.3.2 and Section 5.3.3, we lifted and designed parallel algorithm skeletons for massive remote sensing data processing algorithms with different computation modes. And the recurring patterns of parallel computation and communication in skeletons are pre-implemented on the multi-core cluster with low-level message passing API MPI. These pre-implementations include task partition strategies and data accessing modes consistent with different dependences between computation and data, parallel computing architecture dealing with multi-level parallelism of both inter-node and intra-node, and process synchronization. When a generic parallel skeleton is instantiated and declared, the computations on distributed remote sensing data objects are performed. Firstly, the task would be divided into subtasks by a two-stage task partition strategy, first nodes then intra-nodes, which is consistent with the data partition strategy. Then it would actually load the data blocks owned by each node concurrently through the parallel I/O operations enabled by parallel file system. Finally, the user defined remote sensing sequential code encapsulated in job class would be implemented in parallel by each process. In this situation, the ease of parallel programming could be offered with a minimum concern for architecture-specific parallel implementation behaviors.

5.4.1 Generic algorithm skeletons for remote sensing applications

The remote sensing image data are featured by the geometric and multi-band structures which result in the inherent data parallelism of the remote sensing processing algorithms. There exists a wide variety of remote sensing data processing algorithms, including common remote sensing image processing algorithms and remote sensing information extraction algorithms. These algorithms could be classified into four algorithm categories according to the different dependence between computation and data, pixel-based processing algorithms with data independent computation, neighborhood-based processing algorithms with region dependent computation, band-based processing algorithms with band dependent computation, and global or irregular processing algorithms with global dependent computation.

5.4.1.1 Categories of remote sensing algorithms

Pixel-Based Processing – this category of algorithms refers to those which perform computation on a single pixel and without the request for context. Typically, the computations of these algorithms are data independent and are of excellent data parallelism. This category includes arithmetic or logical operation on pixels of an image, radiometric correction and pixel-based classification like maximum likelihood or SVM classifiers. Assume that S for source RS image, R for result RS image, x and y represent the geometric position of pixel P in the image, b for the band number, f() for the computation performed on pixel P. Then the computation model of this category can be expressed as:

$$R_{b,x,y} = f(S_{b,x,y}) \tag{5.1}$$

Neighborhood-Based Processing: this category of algorithms refers to those which use the corresponding pixel and its close neighbors (regular window) in an input image to calculate the value of a pixel in an output image. This category includes image filter with convolution, image resampling and geometric correction, etc. Here $region(S(b, x, y))$ represents the data region in a neighborhood with pixel P in the source image band(b). The computation model of this category can be expressed as:

$$R_{b,x,y} = f(region(S_{b,x,y}) \tag{5.2}$$

Band-Based Processing: this category of algorithms refers to those that the computation of a single pixel in an output image should use the pixels in the same position or its close neighbors in several input image bands. This category includes pixel-base fusion, image transformation, DNVI, etc. $Vector(S(x, y))$ is for the spectral vector of pixels located in position (x, y) of a multiple image band. Then the computation model of this category can be expressed as:

$$R_{b,x,y} = f(vector(S_{x,y})) \tag{5.3}$$

Global or Irregular Processing: The algorithms in this category are those for which the irregular pixels or even the global image are needed for calculating the value of a single pixel in an output image or statistical features. These algorithms are commonly with poor parallelism. Assume that the representatives for image band (b), computation model of this category can be expressed as:

$$R_{b,x,y} = f(S_b) \tag{5.4}$$

5.4.1.2 Generic RS farm-pipeline skeleton

The farm-pipeline skeleton aims at the remote sensing data processing algorithms with excellent data parallelism, such as pixel-based processing algorithms, neighborhood-based processing and band-based algorithms, but these algorithms must not introduce geometrical warping of the image. Thus this skeleton template is applicable to plenty of remote sensing applications.

Concerning the multi-level parallel architecture of a multi-core cluster, both parallelism among inter-nodes and parallelism in intra-nodes, a farm-pipeline two level parallel pattern is proposed (Figure 5.3) in this algorithm skeleton.

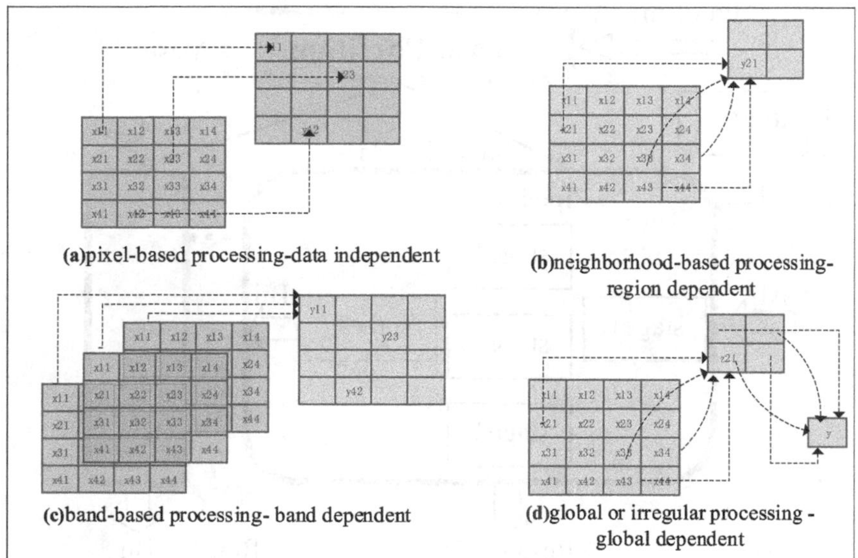

FIGURE 5.3: The algorithm categories and dependences between computation and data.

- First, farm parallel pattern for inter-node. The master node is responsible for the task partition and assignment, also sewing the result data block for output. The worker nodes process the assigned RS data blocks A_i with the user defined function in parallel. In this pattern, the workers could do work dependently without data communication.
- Second, pipeline parallel pattern for intra-node. The data blocks A_i assigned to the node are further split into sub-blocks $A_{i,j}$. And these sub-blocks are then computed by processes of intra-node in parallel. In order to address the time-consuming I/O operations caused by massive remote sensing image data, the idea of "on-the-flow" processing is adopted, the sub-blocks with large scale are loaded into memory and processed as data flows. And the data loading, multiple data processing, and data sewing processes are together to form a processing pipeline. When the pipeline is running, the sub-block $A_{i,j-1}$ is computed by several processes in nodes; simultaneously the sub-block $A_{i,j}$ is pre-fetched by the data loading process. In this condition, the sub-blocks $A_{i,j-2}$, $A_{i,j-1}$, $A_{i,j}$ are processed in parallel. In this way, the intra-node parallelism could be

sufficiently made use of and also the data I/O could be fully overlapped with computation. Figure 5.4 overviews the parallel pattern of farm-pipeline.

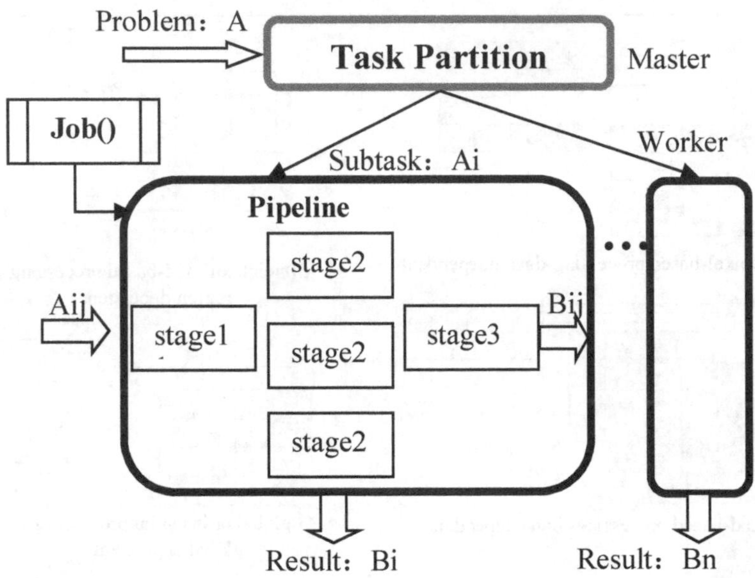

FIGURE 5.4: The parallel pattern of farm-pipeline.

The common sequential processing flow of the skeleton algorithms performing computation on RS data A are expressed as sequence [Load(A), Comp(f(A)), Save(B)]. The load() and save() operations use sequential I/O which is time-consuming. The function f() is applied on each pixel of RS data A, and B is the result of f(A). And the parallel processing flow of this RS farm-pipeline skeleton is expressed as farm $[slice(A), pipeline[load(A_{i,j}), parallel[Comp(f(A_{i,j-1}))], Zip(B_{i,j-2})]]$. The MPI-IO primitives supported by the parallel file system in function load() will be performed by computing nodes in parallel. Thus the parallel I/O is provided which will greatly improve the I/O performance. The function f() is applied on sub-block $A_{i,j-1}$ by several intra-node processes in parallel.

As shown in Figure 5.5, the RS farm-pipeline skeleton is wrapped and implemented as a generic algorithm template class RSFarmPipelineSkeleton with template class mechanism. The parallel architecture, common processing flow and parallel implementation details are pre-implemented in the template. The user-defined sequential code or function is encapsulated in the Job class which provides a virtual function interface *void* operator*() (*void**). As the task partition operation in this skeleton is just a data split operation, it will be done by the constructor of the distributed RS data template.

```
template <class T, class Job>
class RSFarmPipelineSkeleton{
  protected: //definition of template functions
     void load(&RSBlock< T > block);
     RSBlock< T > comp(RSBlock< T > block, Job job);
     zip(RSBlock< T > block);
  public:
     //parallel
     static void doWork(Job job, Dist_RSData<T> dist_A, Dist_RSData<T>
                 &dist_B ){
        //code omitted
        //construct pipeline here with pseudo code
        Stage1: Load (dist_A(i,j));
        Stage2: dist_B(i,j-1)= job (dist_A(i,j-1));
        Stage3: dist_Bi=Zip (dist_B(i,j-2));
        Pipeline(Load (dist_A(i,j)), job (dist_A(i,j-1)),
                 Zip (dist_B(i,j-2)));
  }
};
```

FIGURE 5.5: The generic definition of RS farm-pipeline skeleton.

The API of the RS farm-pipeline skeleton is provided as a static template member functions of RSFarmPipelineSkeleton class (Table 5.1). A class Job is passed as a type argument to skeleton.

TABLE 5.1: API for RS farm-pipeline skeleton.

Constructors of distributed RSData	
Dist_RSData(RSData<T>A, F & policy_f, F & map_f=NULL)	Initialized by the local RSData
API of RS farm-pipeline skeleton template	
static void RSFarmPipelineSkeleton:: doWork(Job job,Dist_RSDatad_A, Dist_RSData & d_B)	Apply job to d_A into d_B

And Figure 5.6 gives an example parallel program for image resampling using the RS Farm-pipeline Skeleton.

```
// data template definition
typedef RSData< unsigned char > rsdata;
//Task Defination
typedef Job < rsdata, rsdata > job;
//skeleton defination
RSFarmPipelineSkeleton< rsdata,job> app;
// Application calling
job resampling;
rsdata A(lines,samples,bands,overlay,"BSQ");
rsdata B=A;
Dist_RSData dist_B(B， policy_f);
Dist_RSData dist_A(A， policy_f);
app:: doWork(resampling,dist_A,dist_B);
B= dist_B.zip();
```

FIGURE 5.6: Image Resampling Program using RS Farm-pipeline Skeleton.

5.4.1.3 Generic RS image-wrapper skeleton

The RS image-wrapper skeleton is designed for the algorithms including pixel-based processing algorithms, neighborhood-based processing and band-based algorithms, the computation of which will causes, not cause the geometrical warping of image data. And also the irregular processing is supported here. These algorithms are also provided with beautiful parallelism. So, the farm-pipeline two-level parallel pattern is also adopted in this skeleton.

These algorithms would be characterized by the computation mode (Figure 5.7) that the calculation of a single pixel in output image requires the irregular region of input image which located in the position determined by a backward geometrical mapping. Typically the input RS image data will follow a formula partition strategy. In this situation, as the sliced data cannot meet the need of calculation, the extra data communications with other nodes will be probably conducted to get data. Consequently, how to express and get the irregular region of data needed before computation and avoiding frequent extra data communications poses a big challenge for writing the parallel programs of these algorithms.

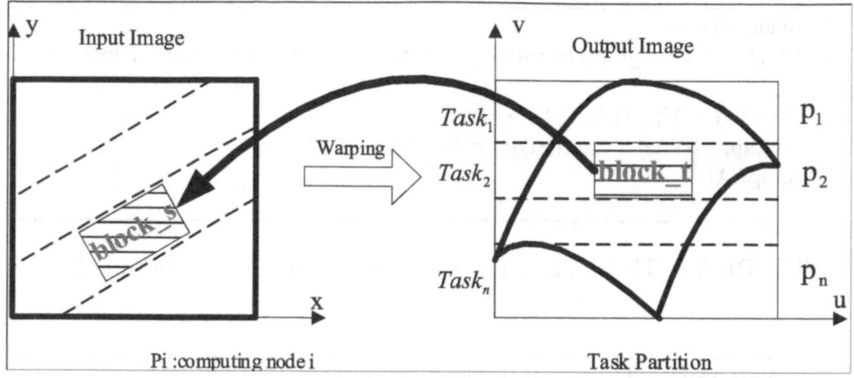

FIGURE 5.7: The computation mode of image warping.

To address the above issues, a default irregular data partition strategy is embedded in this generic algorithm skeleton. The mapf() function defines the geometrical mapping used to calculate the data irregular data region for computing data Bi. And the approximate() function uses the rectangular region Ai to approximate the irregular data region. The function irregularDataSlicing() which implements this strategy is as follows.

The common sequential processing flow of the RS image-wrapper skeleton performing computation on RS data A is expressed as sequence [Load(A), initTask(mapf(A)), Comp(f(A)), Save(B)]. The initTask function uses mapf to compute the image region of the result image B (task initiation) with a backward geometric mapping. The function f() is applied on each pixel of RS data A. And the parallel processing flow of this RS image-wrapper skeleton is expressed as farm[$sequence[initTask(mapf(A)), slice(B)]$, pipeline [irregularDataSlice ($A_{i,j}$),load ($A_{i,j}$), parallel [Comp (f($A_{i,j-1}$))], Zip ($B_{i,j-2}$)]. The irregularDataSlice implements the irregular data partition strategy to slice the image A into blocks which include the irregular data region needed for calculation. The function f() is applied on sub-block $A_{i,j-1}$ by several intra-node processes in parallel.

As shown in Figure 5.9, the RS image-wrapper skeleton is wrapped and implemented as a generic algorithm template class RSImageWrapperSkeleton. The data partition operating on input image A is put off to be done by each worker. Each worker node calculates the exact data block A_i needed for computing the assigned data block B_i with mapf and slices A_i from image A. Thus the frequent data communication incurred by the blind data partition at the beginning of the algorithm could be avoided. The user-defined sequential code or function is encapsulated in the Job class which provides a virtual function interface *void* operator*() (*void**). And the backward geometrical mapping function is wrapped in the MapF class which provides a virtual function interface *void* operator*() (*void**).

```
template<class T >
RSBlock< T > irregularDataSlicing(RSData< T > A ,MapF mapf, Block<T>
Bi ){
   Block<T> Ai; //π(A)={A1, ···An}
   Ai= approximate(mapf(Bi)); //π(B)={B1,···Bn}
   return Ai;
};
```

FIGURE 5.8: The implementation of irregularDataSlicing function.

```
template <class T, class Job, class MapF>
class RSImageWrapperSkeleton {
    protected: //definition of template functions
        void load(&RSBlock< T > block);
        RSBlock< T > comp(RSBlock< T > block, F& f);
        zip(RSBlock< T > block);   public:
    public:
    //parallel
public:
    // Initialize problem with mapf
    static RSData<T> initTask(MapF mapf, RSData<T> A){
        // Initialize problem B with mapf
        RSData<T> B=mapf(A);
        return B;
    }
    // parallel implemented in worker nodes
    static void doWork(F& f, F& mapf, RSData<T> A ,
                              Dist_RSData <T> &B){
        //code omitted
        //construct pipeline here with pseudo code
            Stage1:  A(i,j-1)=irregularDataSlice(A,mapf(),B(i,j-1))
            Stage2:  Load (A(i,j));
            Stage2:  dist_B(i,j-1)=job (A(i,j-1));
            Stage3:  dist_Bi=Zip (dist_B(i,j-2));
            Pipeline(irregularDataSlice(A,mapf(),B(i,j-1)),Load (A(i,j),
                    job (A(i,j-1)), Zip (dist_B(i,j-2)));
    }
};
```

FIGURE 5.9: The generic definition of RS image-wrapper skeleton.

The API of the RS image-wrapper skeleton is provided as static template member functions of the RSImageWrapperSkeleton class (Table 5.2). A class Job is passed as a type argument to the skeleton.

TABLE 5.2: API for RS image-wrapper skeleton.

API of RS farm-pipeline skeleton template	
RSData<T>initTask (MapF mapf, RSData A)	Initialize problem with mapf
static void RSImageWrapper Skeleton:: doWork(Job job, MapF mapf, RSData<T>A , Dist_RSData<T>&d_B)	slice A with mapf , and Apply job to A into d_B

Figure 5.10 gives an example parallel program for fine geometric correlation using the RS image-wrapper skeleton.

```
// data template definition
typedef RSData< unsigned char > rsdata;
typedef RSBlock< unsigned char > rsblock;
//Task Defination
typedef Job < rsdata, rsdata > job;
typedef MapF< rsblock , rsblock > mapFun;
//skeleton defination
RSFarmPipelineSkeleton< rsdata,job, mapFun > app;
// Application calling
job fineCorrection;
mapFun mapf;
rsdata A(lines,samples,bands,overlay,"BSQ");
rsdata B=initTask(mapf(), A);
Dist_RSData dist_B(B, policy_f);
app:: doWork(f,mapf,A,dist_B);
Dist_RSData dist_A(A, policy_f);
app:: doWork(fineCorrection, mapf ,A,dist_B);
B= dist_B.zip();
```

FIGURE 5.10: Image resampling program using RS image-wrapper skeleton.

5.4.1.4 Generic feature abstract skeleton

This skeleton is suitable for the feature abstraction algorithms which perform operations on each pixel to abstract the feature information, and then features would be gathered for global statistical computation. So these algorithms are somewhat global-based processing. The feature abstraction computation can be pixel-based, neighborhood-based or band-based processing following the parallel pattern of the RS farm-pipeline skeleton. This category of algorithms includes histogram estimation, road extraction (polygons or poly-lines yielded by segmentation and detection algorithms) etc. The parallel pattern of this algorithm skeleton is illustrated in Figure 5.11.

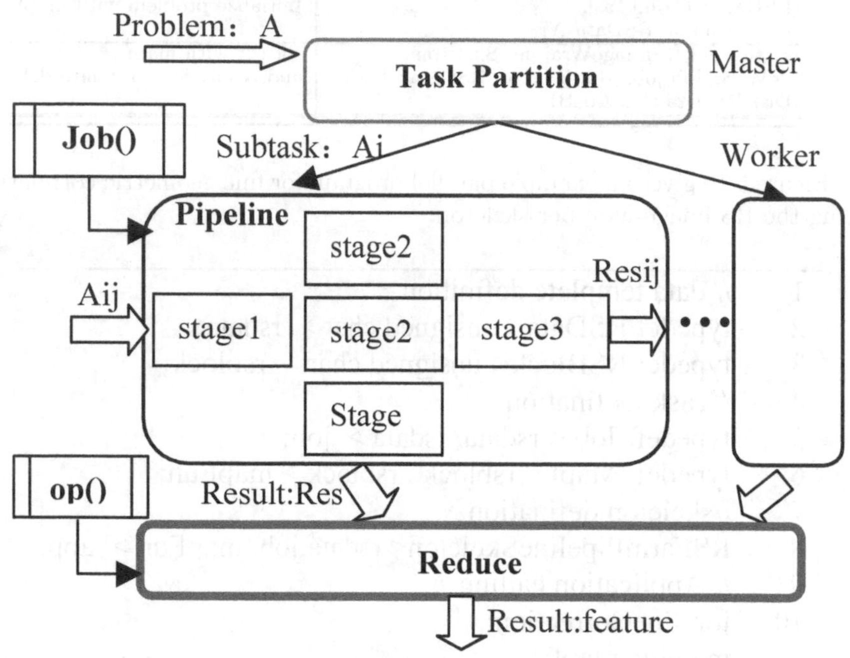

FIGURE 5.11: The parallel pattern for feature abstract skeleton.

The common sequential processing flow of these algorithms performing computation on RS data A are expressed as sequence [Load(A), Comp(f(A)), op(Res), Save(C)]. The function op() if is applied on the staging result of the abstracted features (Res) to compute a final feature of the entire image dataset. And the parallel processing flow of this feature abstract skeleton is expressed as sequence [farm[slice(A), pipeline[load($A_{i,j}$), parallel[comp(f($A_{i,j-1}$))], merge(($Res_{i,j-2}$)]], reduce(op(Res_i))].

The function f() is applied on sub-block $A_{i,j-1}$ by several intra-node processes in parallel to compute res. And the reduce() function performs reduce operation op() on Res_i computed by each node to calculate the final feature.

As shown in Figure 5.12, the RS feature abstract skeleton is wrapped and implemented as a generic algorithm template class RSFeatureAbstractSkeleton. The parallel pattern, common processing flow and parallel implementation details are pre-implemented in the template. The reduce operation function is encapsulated in the OP class which provides a virtual function interface *void* operator() (void*)*.

```
template <class T, class R,class Job, class OP>
class  RSFeatureAbstractSkeleton {
    protected:  //definition of template functions
        void load(&RSBlock< T > block);
        RSBlock< T > comp(RSBlock< T > block, Job job);
        merge(R result);
    public:
        // parallel implemented in worker nodes
        static void doWork(Job job, Dist_RSData<T> dist_A, R&  Res){
            //code omitted
            //construct pipeline here with pseudo code
            Stage1:  Load (dist_A(i,j));
            Stage2:  Res(i,j-1)= job (dist_A(i,j-1));
            Stage3:  Res_i =Merge (Res(i,j-2));
            Pipeline(load(A_{i,j}), comp(f (A_{i,j-1})),merge((Res_{i,j-2})));
        }
        // global reduce operation
        static void reduce(OP op, R  Res, R & Feature){
            //code omitted
            1. define MPI Reduce OP
            2. call MPI_Reduce() operation.
        }
};
```

FIGURE 5.12: The generic definition of RS feature abstract skeleton.

The API of the RS feature abstract skeleton is provided as static template member functions of RSFeatureAbstractSkeleton class (Table 5.3). The class Job and class OP are passed as type arguments to the skeleton.

TABLE 5.3: API for RS feature abstract skeleton.

API of RS farm-pipeline skeleton template	
static void RSFarmPipelineSkeleton:: & Res doWork(Job job, Dist_RSDatad_A, R & Res)	Apply job to A into Res
static void reduce(OP op, R,Res, R & Feature)	Apply Reduce to Res into Feature

Figure 5.13 gives an example of a parallel program for fine geometric correlation using RS feature abstract skeleton.

```
// data template definition
typedef RSData< unsigned char > rsdata;
typedef unsigned char T;
//Task Defination
typedef Job < rsdata, rsdata > job;
typedef OP < R, R > Op;
//skeleton defination
RSFeatureAbstractSkeleton < T,R,job,Op > app;
// Application calling
job histEstimation;
Op op;
R res,feature;
rsdata A(lines,samples,bands,overlay,"BSQ");
Dist_RSData dist_A(A， policy_f);
app:: doWork(histEstimation,dist_A,res);
app:: reduce(op,res,feature);
```

FIGURE 5.13: Histogram estimation program using RS feature abstract skeleton.

5.4.2 Distributed RS data templates

A remote sensing image dataset always consists of complex metadata and multi-band images whose pixels are of multi-dimensionality. The metadata is the data for self-description, which includes image information, geographical information and satellite & sensor information. The geographical information is quite important, as it is closely related to the image data and also involved in the data processing procedure of algorithms. To facilitate the use of the RS data, we abstract the entire remote sensing image dataset with images and metadata into a remote sensing data object named RSData.

5.4.2.1 RSData templates

The RSData object could be mathematically expressed as equation 5.5, where T represents the images with multi-band structure, which stores the spectral information and geometrical location of each pixel. And P is for all the metadata.

$$A = T \oplus P \tag{5.5}$$

The multi-band images of RSData could be expressed with a three-dimensional matrix, for an RSData the image data of which has k band of images in size of m rows and n samples. Then T could be expressed as equation 5.6.

$$T = \begin{bmatrix} a_{111} & \cdots & a_{11n} \\ \vdots & \ddots & \vdots \\ a_{1m1} & \cdots & a_{1mn} \end{bmatrix} \begin{bmatrix} a_{211} & \cdots & a_{21n} \\ \vdots & \ddots & \vdots \\ a_{2m1} & \cdots & a_{2mn} \end{bmatrix} \cdots \begin{bmatrix} a_{k11} & \cdots & a_{k1n} \\ \vdots & \ddots & \vdots \\ a_{km1} & \cdots & a_{kmn} \end{bmatrix} \tag{5.6}$$

Normally, the image data T would be ordered in three sequences, including BSQ, BIL and BIP. As showed in equation 5.6, x is of row dimension, y is of sample dimension and z is of spectral band dimension. The image data $[a_{z,x,y}]$ are ordered in BSQ (band major sequence) as (band, (row, (sample))). The image data $[a_{x,y,z}]$ are ordered in BIP (band interleaved by pixel) as (row, (sample, (band))). The images organized in this order are suitable for band-based processing which performs on the Spectral vector. The image $[a_{x,z,y}]$ data are ordered in BIL (band interleaved by line) as (row, (band, (sample))).

$$T = [a_{z,x,y}] or [a_{x,y,z}] or [a_{z,x,y}], 0 < z < k, 0 < x < m, 0 < y < m \tag{5.7}$$

The metadata P includes image information, geographical information and satellite & sensor information. The image information contains the image size, data type, data arranged order etc., while the geometrical information includes the latitude and longitude of the four image corners in geographic coordinate system, the x and y of the four image corners in the projected coordinate system, and also the projection parameters and the ground resolution of the pixel. In addition, the satellite and sensor information includes the Satellite orbital parameters and sensor parameters. Regarding the RS data from different satellites or sensors, the sensor parameters vary and metadata are always organized in different forms. Also the parameters of different projection methods differ from each other. It means that the data structure of metadata is dynamic which make the expression rather difficult. Thus, we chose the standard WKT projection string of GDAL to express the projection parameter and also the similar string would be provided for expressing the sensor parameter, in order to adapt to different images from different sensors. The RS data object is wrapped as a generic algorithm template class RSData shown in Figure 5.14.

```
template <class T, string satelliteSensor>
class RSData {
    protected: //definition of template functions
        // image data(T)
        T data*;
        //property data(P)
        imageInfo imageInfo;
        geoInfo geoInfo;
        senserInfo senserInfo;
        satelliteInfo satInfo;
    public:
        // constructor
        RSData(long lines,long samples,int bands,string
                dataSequence=”BSQ”)
        {/*initialization here*/}
        // overload operator () + = operations
        inline T operator()(long k,long x, long y) const{
            if (dataSequence==”BSQ”)
                return data[k* rows *samples+ x *rows+j];
            ...}
        //method for getting/setting property data
        string getMeta(string infoName)
        {/*get meta data here*/    }
        //get the latlong value of each pixel
        Latlong latlong(long x, long y)
        {//calculate the latitude and longitude with projection parameter}
        //serialization
        char* msg seriziaiton(){//serialized here}
        deseriziaiton(char* msg){//deserialized here}
        // other interface omitted
};
```

FIGURE 5.14: The definition of RSData template.

The instantiation of RSData template is shown in Figure 5.15.

```
typedef RSData<unsigned char, “spot4”> spot4Data;
spot4Data   s4(lines,samples,bands,”BSQ”);
char pixel=s4 (k,x,y);
string projStr=s4.getMeta ( “projStr” );//get projection parameters
latlong  pos= s4.latlong (x,y); // get the latlong value of each pixel
```

FIGURE 5.15: Instantiation of RSData template.

5.4.2.2 Dist_RSData templates

A remote sensing image dataset whose sliced data blocks are scattered among nodes is abstracted and wrapped by the generic distributed RS data class named Dist_RSData. Programmers do not need to be concerned about the details of data partition and node mapping; the distributed RS data could be operated as a local RS data just like RSData.

The distributed RS data object could be mathematically expressed as equation 5.8, where A is for a normal RS data, $\pi(A)$ is a set of sliced blocks, and map $(\pi(A))$is the set of node mapping of blocks.

$$\begin{aligned} dist_A &= A \oplus \pi(A) \oplus map(\pi(A)),\ A = T \oplus P \\ \pi(A) &= \{A_1, \cdots A_n\},\ map(\pi(A)) = mapf\,\{A_1, \cdots A_n\} \end{aligned} \tag{5.8}$$

Data slicing is used by the Dist_RSData template to recursively slice the remote sensing image data into blocks with a user defined data partition strategy which is suited for relevant algorithms. When the data is sliced, the geometrical information of the sliced block should be recalculated. Assume that an RS image dataset $A(A \in RSData)$, $\pi(A)$ is one partition of A; A_i and A_j are two sliced data blocks adjacent to each other. When the operation of sets is performed like $A_i \bigcap A_j$, the overlay of the blocks are ignored. We offer both the regular and irregular data slicing operation. Where the regular data slicing is a strict data partition method, the sliced data blocks have no intersections. But the data sliced by the irregular data partition method are of irregular shape and they are probably intersected. Then the regular data slicing operation could be expressed as equation 5.9; the irregular data slicing will be expressed as equation 5.10.

$$\pi(A) = \{A_i | A_i \in A,\ and\ A_i \neq \varnothing\},\ \cup_{A_i \in A} A_i = A, i \neq j \leftrightarrow I_i \cap I_j = \varnothing \tag{5.9}$$

$$\pi(A) = \{A_i | A_i \in A,\ andA_i \neq \varnothing\},\ \cup_{A_i \in A} A_i = A \tag{5.10}$$

Here, a two-stage data slicing is adopted in Dist_RSData templates. As described in Figure 5.16, the first stage is for slicing into blocks which would be map inter-nodes and the second stage is for slicing into sub-blocks which would be processed intra-nodes.

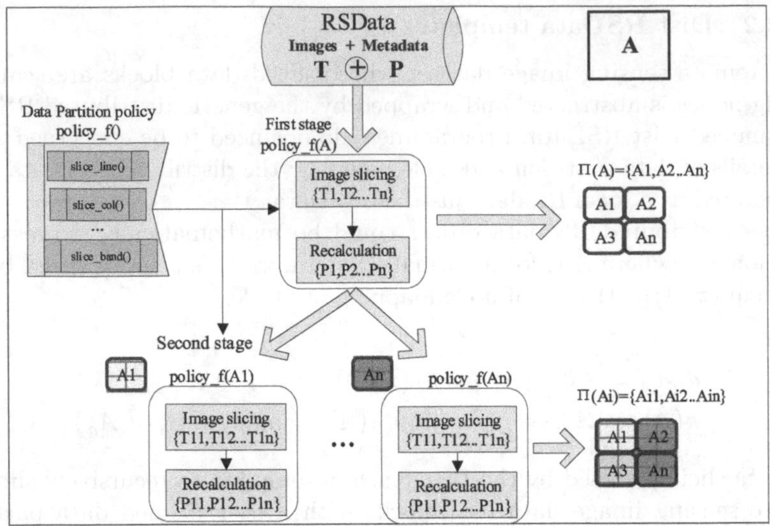

FIGURE 5.16: The processing flow of data slicing.

The dataSlice function template for Dist_RSData templates is abstracted as follows.

```
Template <class T>
array<RSBlock< T >> dataSlice (F & policy_f, RSData<T> A ) {
    // first stage, policy_f (A) -> π (A)={Aij| 0 < i < n, 0 < j < m}
    For(i=0;i<n;i++){
        Ti=Policy_f (T);    // {T1, T2 ... Tn}
        overlay (Ti) ;      //dealing with the overly of Ti
        recalculation(Pi);  // recalculation of Pi
        Ai=Ti ⊕ Pi;
        //second stage: policy_f (Ai) -> π (Ai) = {Ai1, Ai2 ... Aim}
        dataSlicing (policy_f , Ai);
    }
};
```

FIGURE 5.17: The definition of dataSlice function template.

Data distribution, namely the node mapping of sliced data blocks which could be expressed as $map(\pi(A))$ is distributed among $node_1 \sim node_m$. The $< A_i, node_j >, < A_k, node_p >$ is used to represent the mapping relation. Then we can infer that:

$$map(\pi(A)) = \{< A_i, node_j > |0 < i < n,\ o < j < m\} \tag{5.11}$$

$$if(A_i = A_k) \rightarrow then(node_j = node_p) \tag{5.12}$$

Figure 5.18 overviews the data distribution of distributed RS data according to the I/O node mapping of physical data blocks provided by parallel file system when the user defined data distribution function is default. Under this situation, the sliced logical data block is mapping to the computing nodes where the physical data block is located. Thereby, the computation is allowed to be as close as possible to the data, as the data needed for computing is already stored locally. Actually, there exists the situation that the data partition of the logical image and the partition of the physical image in the parallel file system are inconsistent. These means that the logically sliced data block A_i could probably be stored among multiple I/O nodes of the parallel file system. In this condition, the data block A_i would be mapped to the node which contains the most physical data determined by mac($rect_{node_i}$). Thereafter, through data serialization, the sliced data blocks with complex metadata could be resided in local memory of the mapped node. And the region memory space of each node which resides the data block is exposed to all nodes for sharing by one-sided messaging primitives MPI_Win_create and could be virtually converged to global distributed RS data. And all processes could access any block of the distributed RS data as if it is local through RMA (remote memory access) primitives MPI_Get and MPI_Put.

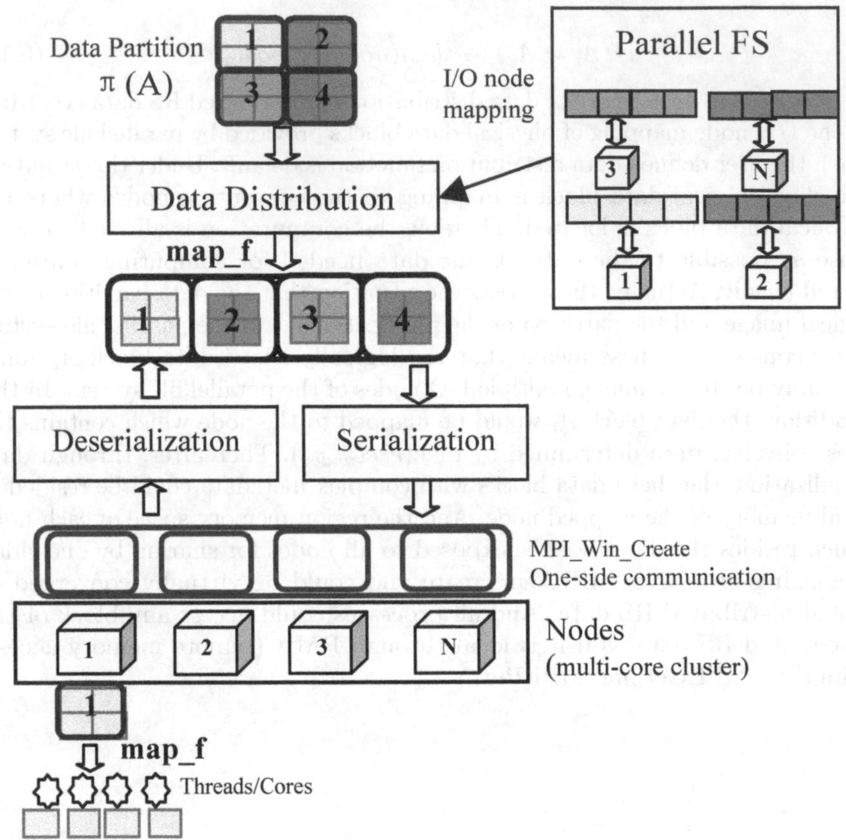

FIGURE 5.18: The processing flow of data distribution.

Data zipping is provided to gather and sew the distributed RS data block. The data zipping operation includes image sewing and metadata recalculation. It could be treated as an inverse processing of data slicing. Before image sewing, the data blocks should be arranged in partial order with the geometrical or spatial position.

```
Template <class T>
RSData<T> dataZip (Array<RSBlock<T>> blockList)
//array(RSBlock):π(A)={A₁, ··· Aₙ}
{
    //sort in band dimention
    blockList =quick_sort_band(blockList)      //sort(Aᵢ ... Aⱼ)
    //sort in row dimention
    blockList =quick_sort_row(blockList)       //sort(Aₖ...Aₘ)
    //metadata merging
    P=merge(blockList);                 //recalculate geometrical information
    T=Sewing(blockList);        //sewing the arrange image block into a image
    A=T ⊕ P
};
```

FIGURE 5.19: The definition of dataZip function template.

5.5 Experiments and Discussion

Generic parallel programming skeletons RS-GPPS have been successfully employed in the parallel image processing system for remote sensing which is short for PIPS. And dozens of remote sensing data processing algorithms are easily parallelized with these parallel programming skeletons and integrated into the system, such as fine correction, NDVI, image registration, fire detection and image classification, etc. The fine correction algorithm was implemented on bj-1 panchromatic image data with size of 48343×19058. It uses CC(Cubic Convolution) resample mode, MQ model and cubic polynomial mapping. The implemented result of this algorithm is shown in Figure 5.20.

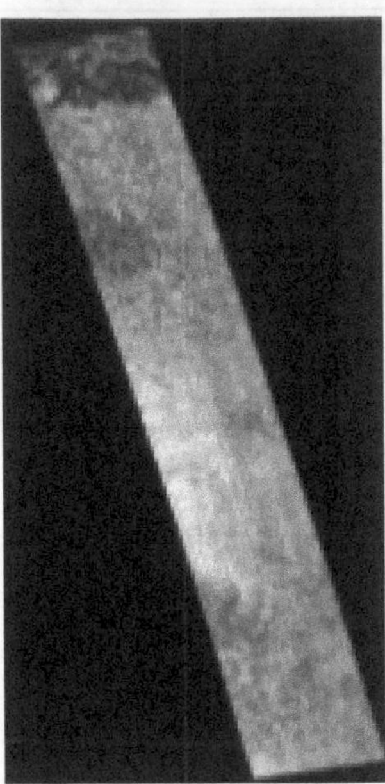

FIGURE 5.20: The result of fine correction algorithm.

RS-GPPS, the generic parallel programming skeletons proposed in this study, lift and wrap generic algorithm skeletons for remote sensing data processing problems and distributed data type templates for RS data. These skeletons offer excellent programmability. To measure ease of programming is anything but easy. The SLOC[152](source lines of code, the comments and empty lines excluded) offers a quantitative metrics for estimation. When programming with parallel skeletons, the extra codes generated by instantiating generic algorithm skeletons are less than twenty lines as shown in Section 5.3. And it is easy programming and would just take dozens of minutes. But when without skeletons, the extra codes generated by implementing parallel operations with MPI would be hundreds of lines. And programming in this way would take great efforts of about a few hours for experts. Two algorithms fine correction (RS image-wrapper skeleton) and radiation correction (RS pipe-line skeleton) which are implemented both with and without skeletons are have been experimented with. The experimental results are listed in Table 5.4.

TABLE 5.4: MSLOC of parallel algorithms.

	Algorithms (With Skeletons)	Algorithms (Without Skeletons)
Fine Correction	386	1414
Radiation Correction	82	554

Communication of remote sensing image datasets with complex metadata structure across nodes is rather trivial and inefficient. Traditionally, plenty of metadata are transferred item by item. In this way, the transferring of one RS data would incur repeated MPI communication operations. Fortunately, we provide serialization operations in data type templates of RS data to serialize the complex metadata structure into a string of characters. Accordingly, when the data is serialized, the transferring of RS data would be done by calling the MPI communication operation only one time. Assume that the metadata contains n data items, the overhead for data serialization is t_s, and mean overhead for transferring one data item is about t_c seconds. Therefore the time overhead for transferring in the traditional way would be $n * t_C$, but if it is done in our way, the overhead would just be $t_s + t_c$. So through data serialization, not only the data communication operation of RS data is simplified but the time used for data transferring would be greatly reduced.

The performance experiments were conducted on a multi-core cluster with 12 multi-core nodes connected by a 20 gigabyte Infiniband using RDMA protocol. Each node is a blade server with dual Intel(R) Quad core CPU (3.0 GHz) and 8GB memory. The operating system was Cent OS5.0, the C++ compiler was the Intel C/C++ Compiler with optimizing level O3, and the MPI implementation was Intel MPI. The remote sensing data processed by algorithms is a bj-1 panchromatic image data with size of 48343 × 19058.

The performances of fine correction (FC) and radiation correction (RC) which were implemented both with and without skeletons (RC-n, FC-n) have been experimented with respectively. The experimental results with the number of computing nodes scaled from 1 to 10 are listed in Table 5.5. And the run time and speedup performance metrics are illustrated in Figure 5.21 and 5.22 respectively. From the experimental result we can see that the time overhead of the algorithms which were implemented with the skeleton are increased by less than 5% when compared to those manually implemented ones.

TABLE 5.5: Performance of parallel algorithms (change node scale).

Nodes (Cores)	FC (s)	FC-n (s)	RC (s)	RC-n (s)
1(1)	696.708	672.46	54.151	53.371
1(8)	83.374	80.539	8.555	8.35
2(16)	37.982	36.679	7.818	7.472
3(24)	29.065	27.739	6.929	6.613
4(32)	20.775	20.064	6.42	6.072
5(40)	18.469	17.723	7.095	6.875
6(48)	15.796	15.161	8.657	8.233
7(56)	17.903	17.312	8.425	7.947
8(64)	13.667	13.079	8.658	8.474
9(72)	17.288	16.631	8.262	7.884
10(80)	19.4	18.605	8.683	8.258

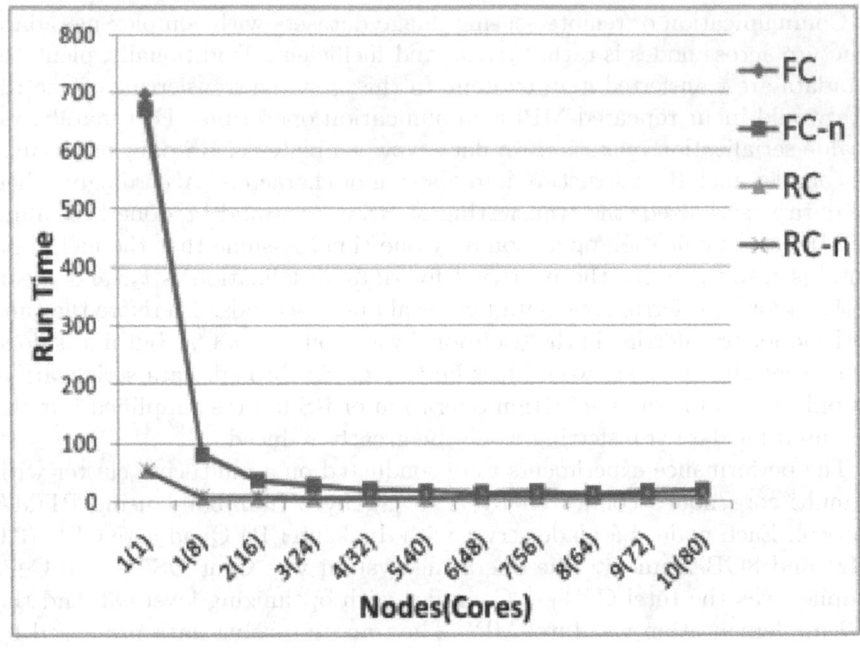

FIGURE 5.21: Run time of algorithms.

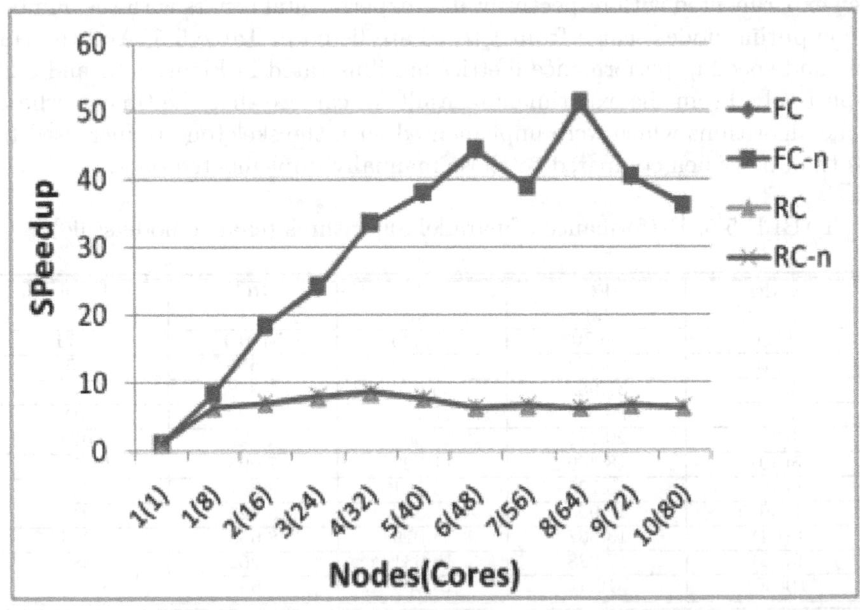

FIGURE 5.22: Speedup of algorithms.

With the amount of RS data increased from 1.3 GB to 507 GB, the experimental results of the above algorithms FC and RC are listed in Table 5.6. The run time and speedup performance metrics are illustrated in Figure 5.21 and 5.22 respectively. These algorithms are implemented with 2 nodes (6 cores).

TABLE 5.6: Performance of parallel algorithms (change node scale).

Data(GB)	FC(s)	RC(s)
1.3	71	5.518411
6.5	250	18.84809
13	203	12.54467
26	393	27.62314
52	730	55.75149
78	1079	78.34589
104	1501	106.0157
130	1842	132.5196
260	3581	273.2002
507	7209	543.991

From the experimental result shown in Figure 5.23, we can know that the programs written with skeletons show excellent scalability with an increasing amount of data.

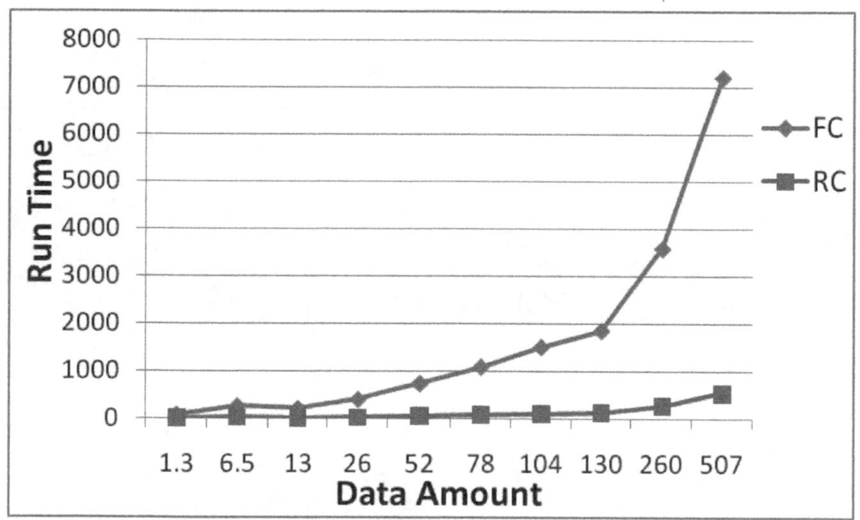

FIGURE 5.23: Performance of algorithms with increasing data amount.

5.6 Conclusions

Parallel programming for data-intensive applications like massive remote sensing processing is quite challenging. As described in this chapter, RS_GPPS is proposed to provide both a template for distributed RS data and generic parallel skeletons for RS algorithms. The distributed RS data template offers an easy and efficient way to distribute the massive remote sensing image data into the exposed memory of each node and also communicate it among nodes. The recurring patterns of computation and communication are pre-implemented in generic parallel skeletons; only by template instantiation could the programs be written as sequential ones with a minimum concern for architecture-specific parallel behaviors. The experimental result shows that SLOC metrics of skeleton-based programs are greatly reduced, with the overhead penalty caused by the template less than 5% compared to manual implantation versions. In addition, these skeleton-based programs also perform excellent scalability with increasing massive amounts of data. The conclusion is that the generic parallel skeletons provided in this chapter is productive and efficient.

Chapter 6

Construction and Management of Remote Sensing Production Infrastructures across Multiple Satellite Data Centers

6.1 Introduction

In recent decades, with the rapid development of earth observing technology, many countries and regions have generally established various satellite platform and satellite data centers with the space observing capacity of the multi-spectral, multi-angle [153], multi-temporal [154] and multi-spatial resolution [155]. And these different types of satellite platforms have generated and will continue to produce vast amounts of remote sensing data. These data will be used to meet the requirements of specialized information extraction and analysis. For example, the United States Earth Observing System Data and Information System (EOSDIS) currently has 12 satellite data centers [156], and its archive data [157] is about 7,000 unique datasets. The total data amount is over 7.5 PB. China RS Satellite Ground Station has four receiving stations which receive the China Brazil Earth Resources Satellite (CBERS), HuanJing (HJ) satellites, Land Satellite (Landsat), SPOT satellites and other data types. The amount of their daily data is about 996 GB, and of their total annual data is about 354TB [158].

Due to a number of factors' limits, such as the revisit period, coverage limits, spectral channels, etc., it is difficult for single-source satellite data to meet the needs of a large-scale integrated RS application. For example, the worldwide production of the 1KM land surface temperature (LST) product needs 4 types of RS data from Advanced Very High Resolution Radiometer (AVHRR), Moderate Resolution Imaging Spectroradiometer (MODIS), FengYun-3 (FY-3) and Advanced Along-Track Scanning Radiometer (AATSR). Therefore, the current multiple satellite data centers agencies should be combined together to provide the services of multi-source RS data. And there is no doubt that carrying out large-scale RS data processing and analysis to meet the demand of different end users has become a popular development trend.

However, there are many problems in integrating multiple satellite remote sensing data centers and building a distributed data processing system. Firstly, a comprehensive remote sensing application requires massive RS data. Due to the multi-source data distributed in different data centers, large-scale data migration could be generated easily. It will not only cause a high load of satellite data centers, but also affect the efficiency of the multi-datacenter collaborative process because of the low data transmission efficiency. Secondly, there are different data levels during the collaborative process, including the raw data, intermediate products and final products. The dependencies between data and products are more complex. On one hand, different products may need the same input data; if these data cannot be effectively managed, it would result in the duplication of some data processing, and thus reduce the overall processing efficiency of the system. On the other hand, RS data processing is very complex, including pre-processing, post-processing and other complex processes. For example, there are many differences between the processing tasks with different types of data. Additionally, the integrated remote sensing applications often require the system to automatically complete the bulk of data processing tasks, which can automatically manage the complex process flow and carrying out the relevant processing in a distributed scenario. In summary, the multi-level RS data management and complex processes flow management are two key issues in building an efficient RS data collaborative processing system based on multi-datacenter infrastructure.

In this chapter, we present the design and implementation of a massive remote sensing data production system based on the multiple satellite data centers infrastructure, the Multi-datacenter Cooperative Process System (MDCPS). In order to solve the duplication problem of data request and data processing, we adopt a spatial metadata repositories and distributed grid file system to build a distributed, dynamic remote sensing data caching system. We build the remote sensing image processing repositories and multi-level task orders repositories for decomposition and manage the complex processing flow, and compose some processing workflow templates and heuristic scheduling rules to automatically match and schedule the specific complex processing. Finally, we provide a use case of remote sensing production on several data centers, to show the feasibility of MDCPS in processing multi-source, massive, distributed remote sensing data.

The rest of this chapter is organized as follows: Section 6.2 describes the related work of distributed remote sensing data processing. Section 6.3 presents the MDCPS environment and its infrastructures. Section 6.4 presents the design and implementation of data management and workflow management in MDCPS. Section 6.5 takes a specific remote sensing production process as an example to evaluate the performance of data management and task scheduling in MDCPS. Section 6.6 gives some further discussion to evaluate the MDCPS. Section 6.7 describes the conclusions and future work prospects.

6.2 Related Work

Spatial big data processing usually requires significant computational capabilities. Several studies have attempted to apply parallel computing, distributed computing and cloud computing to speed up the calculation process [159, 160, 161, 162]. In the area of RS data distributed processing, many grid-based distributed systems were built [163], such as Grid Processing on Demand (G-POD), DataGrid , InterGrid, MedioGrid, Earth Observing System Data and Information System (EOSDIS) [164]. In these distributed data processing systems, data management and workflow management are two important components. For the distributed remote sensing data management, data transmission [165] is often carried out by means of grid middleware, such as GridFTP [166]. In the aspect of data access and integration, multi-source data are often exchanged in accordance with certain standards or the corresponding conversion, for example, the data format of Committee on Earth Observation Satellites, and the data standard of Open Geospatial Consortium (OGC) and Geographic information/Geomatics ISO/TC 211 [167, 168]. In the area of replicas management, Globus Tookits middleware are often used for data replication and distribution [169]. Gfarm Grid File System [170], as a distributed file system, designed to support the data-intensive calculation based on wide area network (WAN), combines each local file system to be a global virtual file system through the metadata server MDS and distributed I/O nodes, and improves the read and write bandwidth for distributed replicas. L. Wang, et al. [171] have designed and implemented a distributed multi-datacenter system G-Hadoop, which applies the MapReduce framework [172, 173] to the distributed clusters. In the area of workflow management, many scientific workflow systems have made important progress, with a certain ability in tasks monitoring and control, scheduling policy management, and workflow fault tolerance [174]. But those work management systems are only designed and developed for the specific computing scenarios in their research, and lack the common skills of abstract describing. So it is very difficult to meet the needs of different users when they require the deep customization of scheduling policy, fault tolerance, etc.

The remote sensing data processing system based on multi-datacenter infrastructure is a solution to process massive, multi-source and distributed remote sensing data, and the current research in this field is under development. The system based on multi-datacenter infrastructure can improve the efficiencies of the acquisition, organization and processing of distributed data [175]. Like grid-based distributed processing systems, data management and task scheduling are also two major challenges among the massive spatial data processing under this infrastructure. Most relative studies are focused on the algorithm of task scheduling [176, 177, 178] so far. For example, W. Song, et al. [179] proposed the task scheduling mechanism and framework for spatial information processing and geocomputation across multiple satellite data centers; W. Zhang, et al. [180] proposed a workflow scheduling method based on nearby data calculation, and designed a kind of image processing infrastructure based on a multi-satellite center. However, few studies focus on the aspects of data management across multiple data centers, particularly on the data management of distributed collaborative processing.

6.3 Infrastructures Overview

6.3.1 Target environment

As a research result of the 863 program, MDCPS is designed to produce the large-scale and global coverage RS data production based on multi-datacenter infrastructure. It combines the China Centre for Resources Satellite Data and Application (CRESDA), National Ocean Satellite Application Center (NSOAS), National Satellite Meteorological Center (NSMC), Computer Network Information Center (CNIC), Twenty First Century Aerospace Technology Co. Ltd (21AT), and Institute of Remote Sensing and Digital Earth (RADI). More than 60 kinds of RS data types, which are sourced from the Aqua satellite, Terra satellite, Landsat series,CBERS, ZiYuan (ZY) satellite series, HaiYang

(HY) satellite series, FengYun (FY) satellite series, BeiJing-1 (BJ-1) satellite series, National Oceanic and Atmospheric Administration (NOAA) satellite series, Multi-functional Transport Satellites (MTSAT) series, etc., could be processed in this environment. The amount of RS data could be over 1PB. It aims to provide a safe, reliable and efficient environment to support the applications of massive remote sensing production. Currently, MDCPS has the production capacity of more than 40 kinds of RS products, and its products are summarized in Table 6.1.

We keep the following goals when developing MDCPS:

- Effective management of RS data: In the collaborative process, we try to effectively manage raw data, intermediate products and final products. We try to use the data dependencies and data caches, and try to avoid the large-scale data migration by reducing the duplication of data processing;

- Automated processing platform: Aiming at improving the efficiency of co-processing, we try to implement the following process automation: match the complex process workflow, task decomposition, workflow organization, and workflow scheduling.

6.3.2 MDCPS infrastructures overview

MDCPS adopts the centralized system framework for massive RS data production. It consists of a master datacenter and several different data centers in geographic distribution. Master data center (MDC) is mainly composed of the data management system (DMS) and business processing system (BPS). DMS manages raw data, intermediate products, and final products in co-processing, and provides the service of data query, scheduling and data discovery. BPS is responsible for overall mission receiving, workflow organization, and task scheduling. Each data center consists of two subsystems: one is its own data service system with the responsibility of providing raw data services; the other one is the task execution proxy system (TEPS) with the responsibility for pre-processing raw data. The MDC will decompose the global task, schedule each sub-task to the data center's proxy system over the WAN, post process together after merging the intermediate processing results in MDC, and ultimately complete the processing tasks. The system architecture diagram is shown as Figure 6.1.

The software architecture of MDCPS is shown as Figure 6.2, including the application interface layer, business logic layer, software architecture layer and resource layer. In the resource layer, the distributed resources underlying MDCPS include RS data resources, algorithms resources and computing resources over data centers. In the software services layer, we adopt MyProxy and Globus Simple Certificate Authority (CA) as its security certification middleware between the data centers. And we use the Gfarm grid file system to manage the distributed data replicas, use GridFTP to supply distributed data

TABLE 6.1: The RS products in MCCPS, their spatial and temporal characteristics.

Product ID	Product Name	Spatial Resolution	Temporal Resolution
ADR	Aerodynamic roughness	1km	5d
AOD	Aerosol optical depth	1km	1d
AOD	Aerosol optical depth	30m	10d
ARVI	Atmospherically resistant vegetation index	1km	5d
ARVI	Atmospherically resistant vegetation index	30m	10d
BRDF	Bidirectional reflectance distribution function	1km	
CLI	Cloud index	1km	5d
DLR	Downward longwave radiation	5km	3h
DSR	Downward shortwave radiation	5km	1d
ET	Terrestrial evapotranspiration	25km	5d
EVI	Enhanced vegetation index	1km	5d
EVI	Enhanced vegetation index	30m	10d
FPAR	Fraction of photosynthetically active radiation	1km	5d
FPAR	Fraction of photosynthetically active radiation	30m	10d
FVC	Fractional vegetation cover	30m	10d
FVC	Fractional vegetation cover	1km	5d
HAI	Hydroxy abnormal index	30m	365d
LAI	Leaf area index	1km	5d
LAI	Leaf area index	30m	10d
LHF	Latent heat flux	1km	1d
LSA	Land Surface Albedo	30m	16d
LSA	Land surface albedo	1km	5d
LSE	Land surface emissivity	1km	1d
LST	Land surface temperature	1km	1d
LST	Land surface temperature	5km	1d
LST	Land surface temperature	300m	4d
NDVI	Normalized vegetation index	1km	5d
NDVI	Normalized vegetation index	30m	10d
NDWI	Normalized difference water index	1km	1d
NPP	Net primary productive force	1km	5d
NPP	Net primary productive force	300m	10d
NRD	Net radiation data	300m	4d
PAR	Photosynthetically active radiation	5km	1d
PRE	Precipitation	10km	1d
SAI	Suicide abnormal index	30m	365d
SBI	Soil brightness index	1km	
SBI	Soil brightness index	30m	
SHF	Sensible heat flux	1km	1d
SID	Sea ice distribution	1km	10d
SMI	Soil moisture index	1km	1d
SWE	Snow water equivalent	25km	5d
TCWV	Total column water vapour	1km	1d

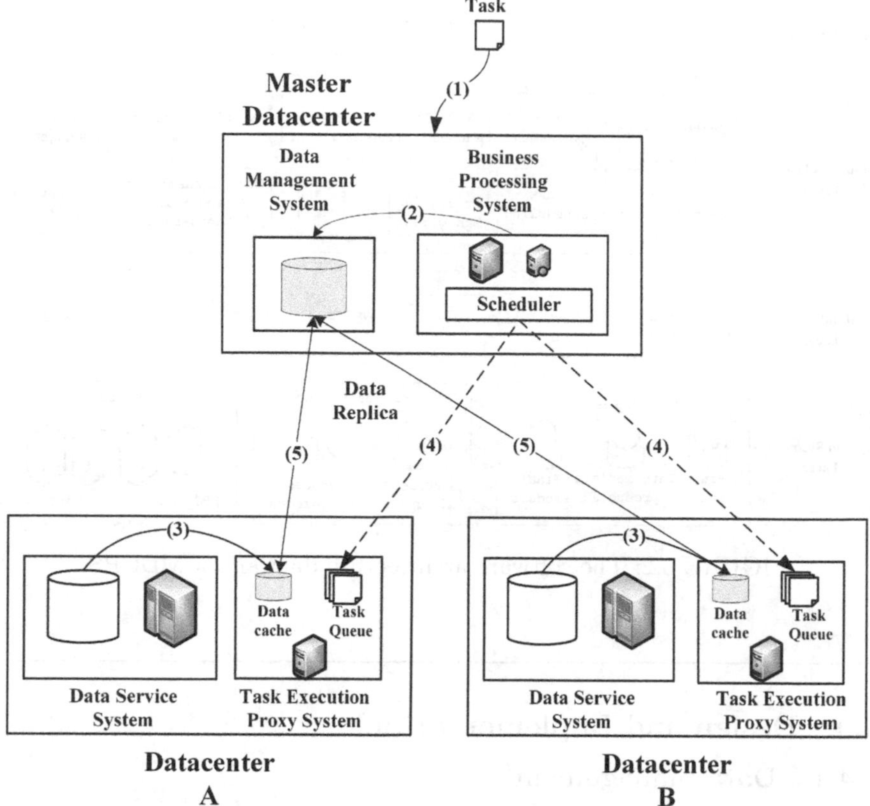

FIGURE 6.1: The system architecture diagram of MDCPS.

transmission services, use Ganglia to monitor the TEPS on distributed satellite data centers and get the information of performance. At the same time, the Kepler scientific workflow system is chosen as our processing workflow engine and we adopt MySQL as the backend database to complete the persistence of all data. In the business logic layer, MDCPS has the daemon module, data management module, workflow management module, task scheduling module, order management module, spatial metadata management module, computing resource management module, algorithm management module, log management module and other functional modules. These modules are used to manage all the distributed data replicas and organize tasks workflow automatically, scheduling and execution. In the application interface layer, the web portal of MDCPS supplies a friendly interface for users to submit their needs for processing massive RS data.

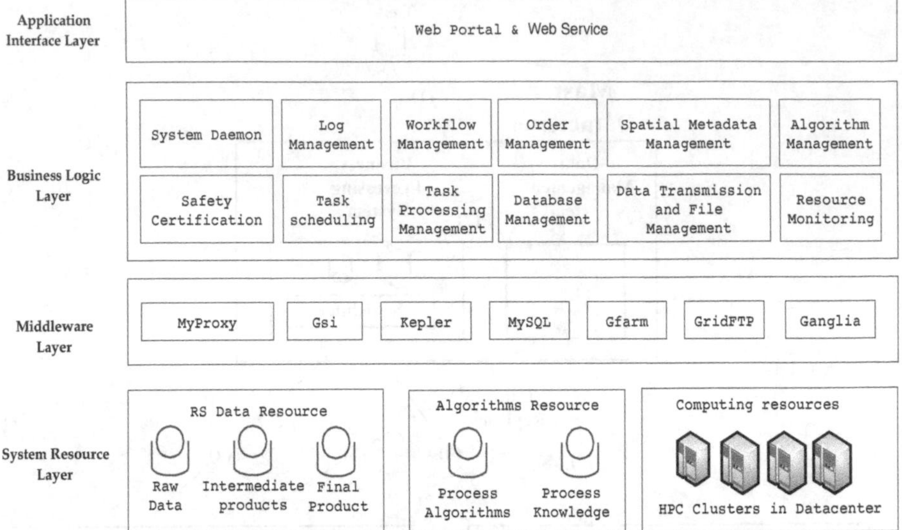

FIGURE 6.2: The software architecture diagram of MDCPS.

6.4 Design and Implementation

6.4.1 Data management

Global and large coverage of RS products often require massive input datasets, and the distributed computing, multi-datacenter collaborative process has huge amounts of data transfer, including raw data, intermediate products and final products. And as a multi-user remote sensing data processing platform, there may be a lot of the same processing requests, the same spatial and temporal scales of input datasets, or the same remote sensing products. All these mentioned above can cause a lot of repeat transmission and processing. To reduce unnecessary duplication transmission and production, MDCPS needs to achieve a unified management of raw data, intermediate products and final products in the process of co-processing. It would not only need to manage the metadata of distributed data, achieve reliable distributed file operations, but also need to manage the complex relationships between RS data. If these issues are addressed, MDCPS would be able to use an effective data scheduling strategy based on the spatial relationships, the distribution of relations, affiliations, dependencies between RS data. It could maximize reuse of the cached data and reduce large scale data migration in the collaborative process.

In MDCPS, we used the strategies of "Spatial Metadata Management & Distributed File Cache Management" to realize the dynamic data management of multi-datacenter collaborative process. Firstly, for the management of metadata and data relationships aspect, we established three basic metadata repositories in MDC to manage the spatial metadata information of the raw data, intermediate products and final products. And then we established a public repository encoding geographic coordinates for the unity of all data spatial relationships. In addition, knowledge repositories of final products were established in MDC for input data parsing for RS products. We also added a series of relationships knowledge between products and implemented the management of products' relations. For distributed data file management, we realized a unified management in the data catalog access, data transmission and data cleaning. In catalog management, we built specific cache directories in MDC and other datacenters. The cache directories in MDC, as a products cache catalog, would cache intermediate products and final products. These directories in TEPS across each data center are used to store the raw data downloaded from the data center. All data files' information about distribution and request will be registered to spatial metadata repositories. Therefore, we achieved unified management of the cached data file's metadata. For the capacity monitoring and cleanup of data cache catalogs, we adopted the Ganglia system client to monitor the capacity of the cache directory in near real-time. If exceeding limited quotas, we would choose specific data by querying spatial metadata repositories. In data transmission, we deployed GridFTP to implement a safe, efficient and stable data transmission service. The MDCPS data management system structure is depicted in Figure 6.3.

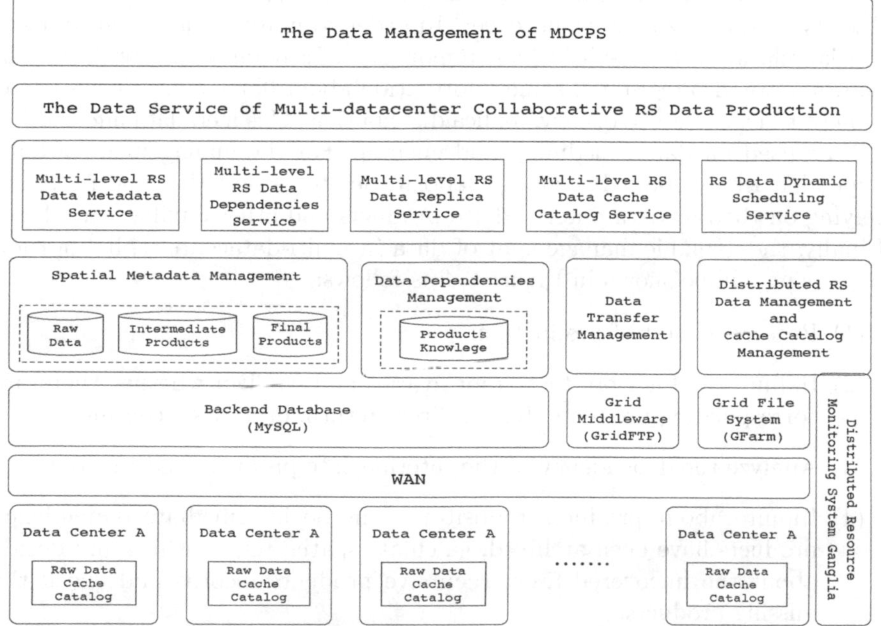

FIGURE 6.3: The data management system of MDCPS.

6.4.1.1 Spatial metadata management for co-processing

A normal production process in an RS data processing system includes: accepting request, parsing the input raw data, raw data query, data download, preprocessing, post processing, products registration to the repositories, and final products feedback. Usually, there are two typical scenarios to meet the user's products need in a processing system.

- Carry out the whole process, and finally feedback processed products;
- Direct feedback of these products, because all the needed products have been produced and archived.

To achieve the goal of greatest reuse of the resource, reduce the large-scale data migration and repetitive processing, and in addition to the above two production scenarios, MDCPS also has the dynamic production scenario based on the cached raw data and intermediate products. In the production process of multi-datacenter co-processing, it not only requires the management of the final products, but also requires the management of the intermediate data, such as the downloaded raw data and the intermediate products after pre-processing (standard products) and other auxiliary products. If realized, we could avoid the duplication of raw data download and intermediate products production.

In order to achieve dynamic management, we built a spatial metadata repository to manage the metadata of raw data, standard products and final products. The metadata information includes data ID, file name, created time, data type, data size and cloud cover. In order to unify management of data spatial relations, we established a common grid of geographical coordinates. In addition, we also registered some information about data replication, such as request time, request frequency, replica distribution. This metadata information will be used for data scheduling and cleanup. For the management of data dependencies, we built knowledge repositories to resolve the final products relying on the raw data, intermediate products and other auxiliary products. Finally, the dynamic management of data in multi-datacenter collaboration processing will be shown in Figure 6.4 as follows:

(1) Receive request of production;

(2) Inquire the final products repositories and confirm whether there are corresponding products. If yes, direct return, otherwise, continue;

(3) Analyze the dependency of the intermediate products and raw data;

(4) Inquire about products repositories, if the intermediate dependency products have been archived, calculate spatial relationships and decide about the uncovered area, recursive products records and check the missing products;

(5) Keep on querying until all the missing products data have been identified and make sure the corresponding data plan is in place;

(6) Query the raw data cache repository to determine whether you need the data downloads;

(7) Query distribution information of data replicas, select an appropriate replica.

(8) Determine the final data plan dynamically.

6.4.1.2 Distributed file management

In the management of the distributed RS data file, we integrated the Gfarm grid file system, GridFTP, Ganglia monitoring system and our spatial metadata repositories. It supplied the services of distributed data management, data transmission, file operations and catalogs monitoring. The efficiency and consistency of distributed file operations could be guaranteed in this environment. Figure 6.5 shows the system deployment.

Firstly, we built a cache catalog in each data center for storing data replica files. A newly created file in this catalog would be uploaded to Gfarm file system. In addition, built a public cache catalog in the master center of MDCPS and used Gfarm2fs to mount the Gfarm system to this cache catalog. Then, all replicas of RS data in the distributed Gfarm system could be displayed in this catalog. Just like local files, we could use the gfexport command to export data easily. Since the metadata in the Gfarm backend database contains limited information and lacks spatial metadata information, we saved the data distribution information in our spatial metadata repository which could instead service replicas metadata in Gfarm. To ensure the consistency of distributed remote sensing data files and its metadata, we updated the corresponding information when it altered. The MDC is responsible for the global control of data transmission between data centers, which requires the third-party control of data transmission. In addition, large-scale data transmission requires a secure and reliable data transfer service. So we adopted the GridFTP as MDCPS's data transmission middleware.

Secondly, with the ongoing production, the cache catalog in each datacenter will keep caching lots of RS data and when it is out of quota, the data transmission and production will terminate. Therefore, it is necessary for MDCPS to adjust and manage cache catalog capacity automatically during production. In MDCPS, the capacity monitor was used, whose specific programs of data replacement and clean-up are as follows: the Ganglia monitoring system was adopted to monitor disk usage. When a datacenter's usage exceeds the quota threshold, the monitor daemon in MDC will throw a warning. Then, MDCPS will use the Gfarm client to statistically determine the data amount in that cache catalog, and the MDC will decide whether to clear the cache catalog according to the feedback. If there is need to clean up, the system will query the metadata information of data replicas, filter some data based on the Least Recently Used (LRU) algorithm and delete these data replicas in the

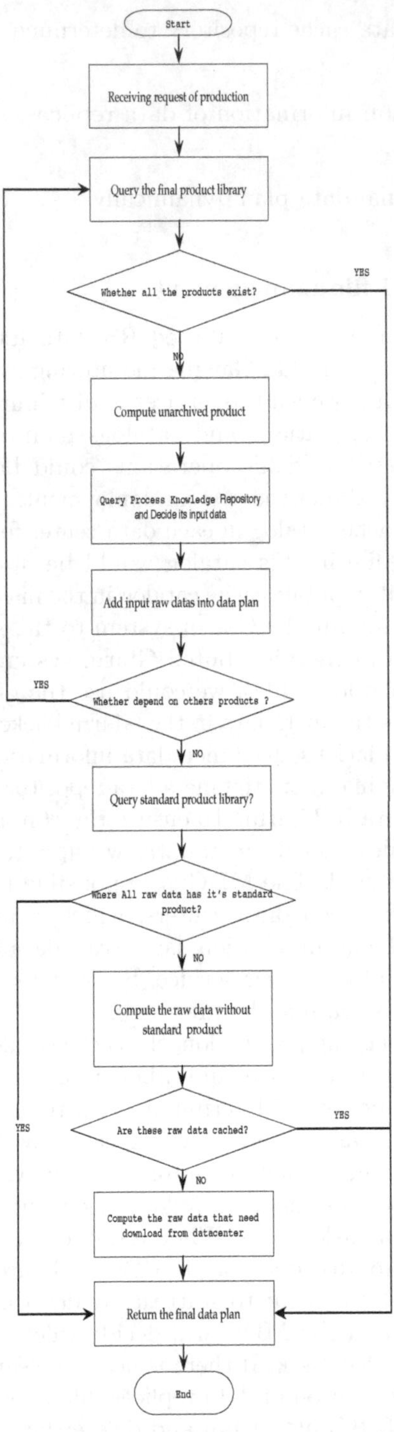

FIGURE 6.4: The RS data dynamic scheduling strategy in MDCPS.

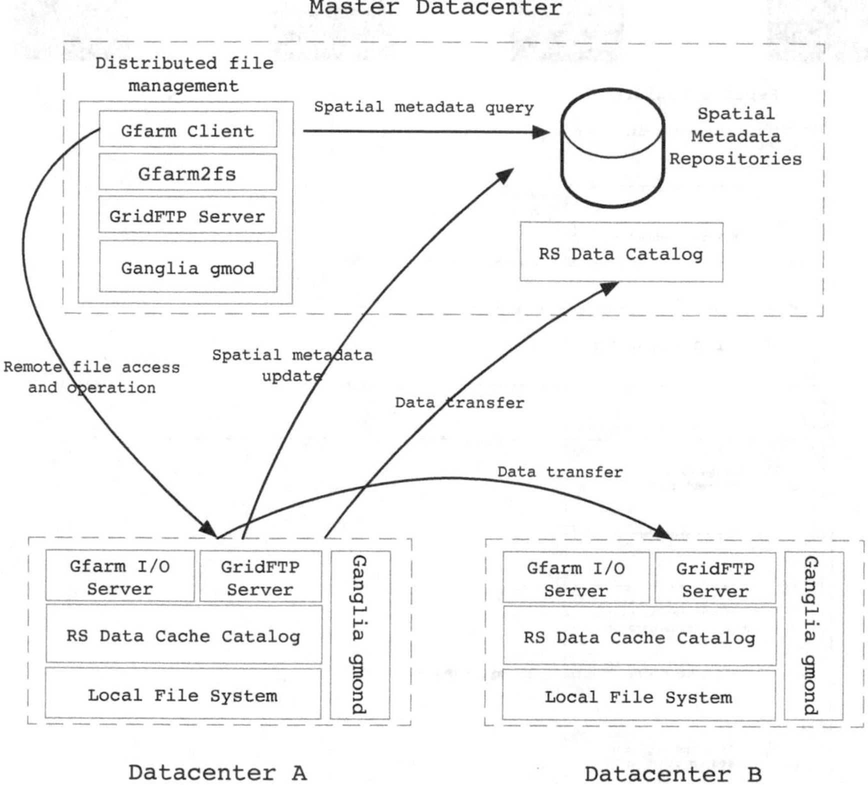

FIGURE 6.5: MDCPS distributed file management.

corresponding cache catalog by Gfarm client. The sequence of data monitoring and cleanup operations is shown in Figure 6.6.

6.4.2 Workflow management

Workflow management is a key module of the distributed processing system, which determines the reliability and stability of the system. The processing tasks of large-scale RS data based on the multi-datacenter infrastructure are complex, and it is difficult to organize, manage and schedule. The difficulties can be classified into two aspects:

- Complex distributed remote sensing data processing: Remote sensing data processing involves pre-processing and post-processing generally, and the processing methods are very different between different kinds of data; what's more, due to the different distribution of data and computing, the input and output data parameters of different processing flow are

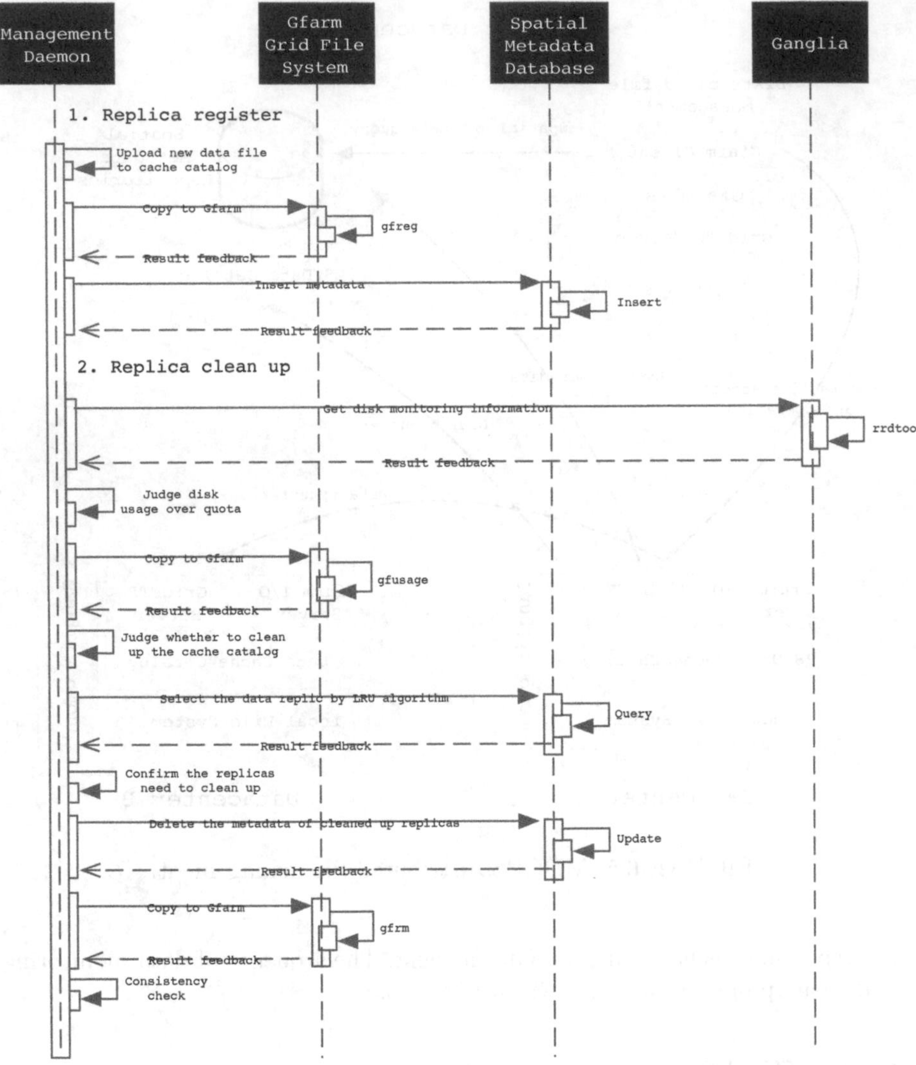

FIGURE 6.6: Distributed file management sequence in MDCPS.

quite complex, and it is difficult to automatically match and organize a suitable workflow in a distributed environment;

- Complex scheduling of multi-datacenter scenarios: The Multi-datacenter collaborative processing system combines multiple data centers to build a system of systems. The single scheduling policy makes it difficult to meet this hierarchical scheduling scenarios. In addition, the task scheduling in MDC involves a variety of dynamic resources across distributed datacenters, including data resources, computing resources and algorithm resources. An optimal scheduling model needs a comprehensive consideration for overall factors and this is a difficulty so far.

To solve the problems above, MDCPS builds an entire workflow system which connects the MDC and other datacenters. In MDC, we constructed a process flow repository and multi-level task orders repositories for matching and decomposition of complex processing tasks. We integrated a Kepler workflow system as the workflow execution engine, and built a Kepler workflow template repository for RS data production. The workflow template could help achieve the automatic construction of concrete workflow. Considering the features of hierarchical architecture, a two-level scheduling strategy was adopted. MDC will schedule tasks to each datacenter by a heuristic scheduling repository. TEPS over each datacenter batch the sub-processing tasks by Torque PBS. The monitoring service of task order could effectively monitor workflow status and provide support for workflow fault-tolerance. The workflow management architecture of MDCPS is depicted in Figure 6.7.

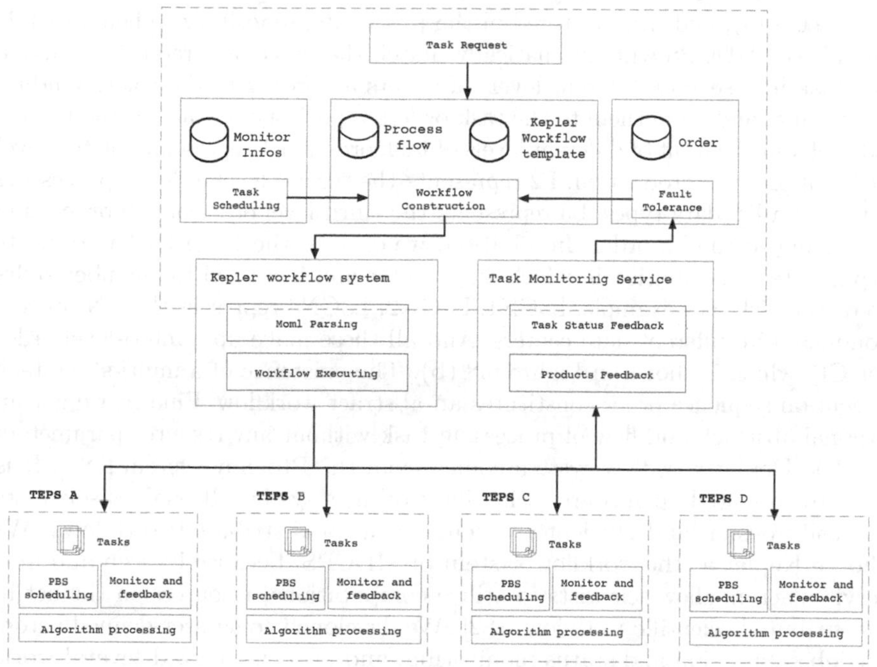

FIGURE 6.7: The workflow management architecture of MDCPS.

6.4.2.1 Workflow construction

In order to solve the complex issues of distributed remote sensing data processing, we constructed a processing repository, multi-level orders repository and workflow template repository to decompose, decouple and map the complex processing task. To begin with, we distinguished the different process flow by unified naming, and established a unique processing depending on its corresponding RS data types. Then, we divided it into several sub-processes following the pre-processing, post-processing steps of each process flow. Finally, we stored the hierarchical relations between different levels of processing flow in the process repository. For example, CP represents a top level processing of common RS products. CP1 and CP2 are two subclasses of CP in the second level. CP1 represents the processing flow of Landsat data and CP2 represents the processing flow of MODIS data. CP1 processing contains five sub-processes in the third level: data preparation (DP), geometric normalization (GN), radiation normalization (RN), standard product uploading (SPU), common product production (CPP) and common product register (CPR). In addition to the former five sub-processes, CP2 also has a data swath (DS) sub-processing. The hierarchical relationship between different CP processing is shown in Figure 6.8 (a), and will be stored in the processing repository. When a task is submitted, MDCPS will automatically match the processing repository, divide the task into several different levels of orders according to the corresponding processing, and store them to the task orders repository. We added the L1, L2, L3 prefix in the front of different levels of task orders. L1 represents the top level order of product production, L2 represents the second level order of processing for certain RS data types, L3 represents the third level order of sub-processing belonging to the L2 order. In addition, we divided the L3 order based on its input data's distribution in which data center and increased the number suffix to distinguish. For example, L3GN1, L3GN2, L3GN3 represent the GN process conducted in different data centers. And all these make up a multi-level order for CP, which is shown in Figure 6.8 (b). The existence of a multi-stage task execution sequence order constitutes an abstract workflow. Finally, we got an original abstract workflow of processing task without any resource parameters.

The Kepler workflow system comes from the Ptolemy system [181]. It is an actor oriented open source scientific workflow system. It enables scientists to easily design and efficiently execute local or distributed workflows. We choose Kepler as the workflow system of MDCPS, because its web and grid services actors allow scientists to utilize computational resources on the net in a distributed scientific workflow [182]. We developed some user-defined actors for job submission and status monitoring, and also customized several workflow templates corresponding to RS data processing based Kepler's Modeling markup language (Moml). Based on these templates, abstract workflow will get specific data, algorithms, and computing resources after workflow scheduling, and become a concrete workflow automatically. The workflow organizational process is shown in Figure 6.9.

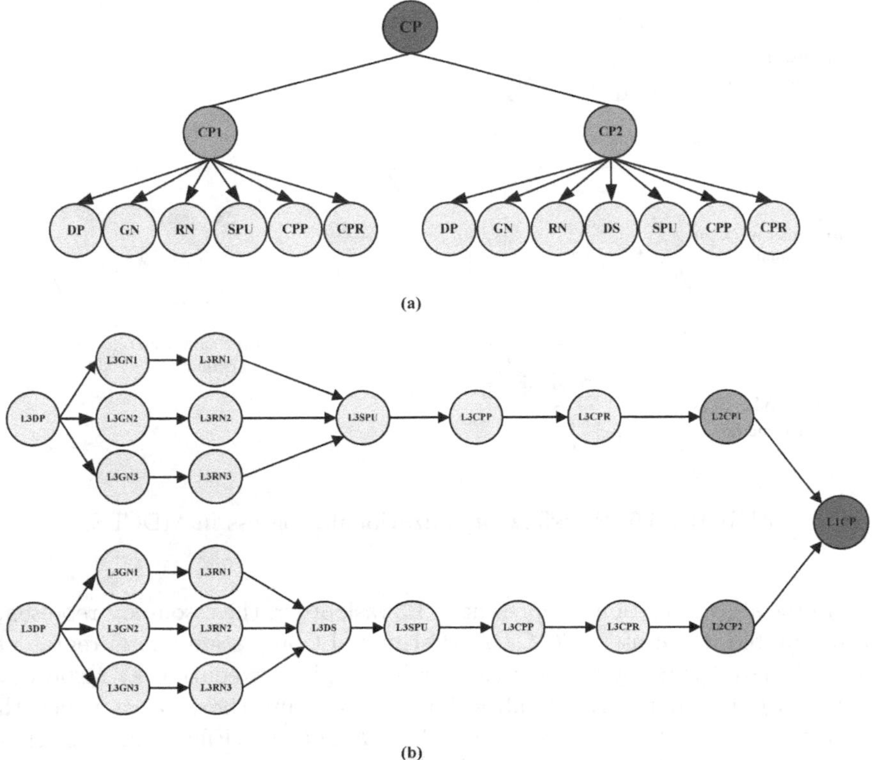

FIGURE 6.8: The processing tree and multi-level task orders.

6.4.2.2 Task scheduling

The multi-datacenter collaborative process is a process of data-intensive computing. MDCPS system task scheduling strategy should consider not only the performance of distributed computing resources, but also consider large-scale data migration. We investigated the correlation scheduling algorithm, and finally applied Min-Min algorithm [183] to our system scheduling strategy. The difference between Min-Min scheduling and our scheduling is that we choose the computing resource (data center) instead of computing tasks in this step. Based on Best-effort scheduling, MDCPS focuses on the execution time constraints to build the objective function. The target of scheduling is to achieve a minimum execution time for L3 task orders, and scheduling objective function is shown as Equation 6.1:

$$ECT(t,r) = max\{EAT(t,r), FAT(t,r)\} + EET(t,r) \tag{6.1}$$

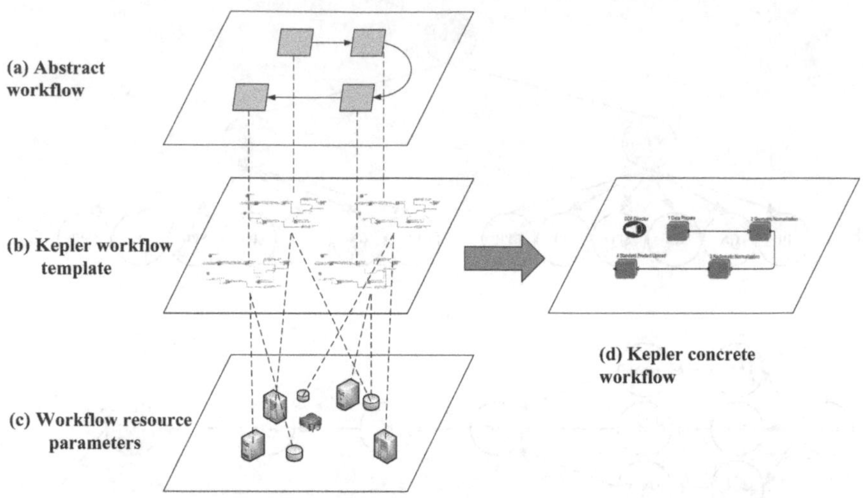

FIGURE 6.9: Workflow organizational progress in MDCPS.

In the above equation, t represents a L3 task order, the resource r represents a distributed datacenter. $ECT(t, r)$ (Estimated Completion Time) represents the estimated time by which task t will complete execution at resource r. $EAT(t, r)$ (Estimated Availability Time) represents the time at which the resource r is available to execute task t. $FAT(t, r)$ (File Available Time) represents the earliest time that all the required RS data files of the task t will be available at the datacenter r. $EET(t, r)$ (Estimated Execution Time) represents the amount of time the datacenter r will take to execute the task t, from the time the task starts to execute. We computed ECTs of each task on all available datacenters and obtained the MCT (Minimum Estimated Completion Time) for each task. We assigned the task on the datacenter to complete it at the earliest time. The basic steps of this scheduling are listed in Algorithm 1 as following.

In the data scheduling and computing resource scheduling stage, the time estimation methods are shown as follows:

In the stage of RS data scheduling, the RS data that have been cached in the MDCPS spatial metadata repository will be the first choice to allocate. For non-cached data resources, the system sends a download request to the corresponding data center according to the data type. We will estimate each datacenter's $FAT(t, r)$ according to the amount of the data request.

In the stage of computing resource scheduling, MDCPS will dynamically monitor system performance information for each datacenter by deploying the Ganglia system, The information such as CPU, I/O, memory, network, load, will update in near real-time to the computing resource repository. We are able to predict the capacity based on this monitor information of the datacenter in the scheduling stage, and get the $EET(t, r)$. In addition, we can predict

Algorithm 1 Scheduling algorithms based on Min-Min

```
while ∃ task ∈ U is not scheduled do
    priorTask ← get an unscheduled ready task whose priority is highest.
    DOSHEDULE(priorTask)
end while
procedure DOSHEDULE(t)                    ▷ Select the optimal data center
    while task is unscheduled do
        for all rΞ availDatacenters do
            compute ECT(t, r)
        end for
        R ← min{ECT(t, r)} ▷ get a datacenter with minimum ECT(t, r)
        schedule t on R
    end while
end procedure
```

availability time by querying each data center's PBS task queue and task order running status in a task order repository, and estimated the $EAT(t, r)$.

On this basis, we built heuristic scheduling rules repository on the basic principle of "Near Data Calculation". It contains some heuristic rules for scheduling, including some empirical parameters for compute FAT, performance indicators of weight parameters for EET and EAT, performance metrics thresholds for resource scheduling, etc. MDCPS could dispatch the processing tasks automatically to appropriate data centers based on the heuristic scheduling rules.

Take the L3GN order processing for example to explain the scheduling method in MDCPS. Under the assumed conditions that the MDCPS system has five data centers (DC1~DC5), DC2 caches 60% of the required RS data and DC3 caches 20%. The remaining 20% of the required RS data will be required and downloaded from some data centers. When the L3GN order is submitted to the MDCPS scheduling system, the detailed scheduling processes are as follows:

(1) Firstly, MDCPS determines which input data are already cached in the data center based on the data management system, and which data should be requested and downloaded at this time. In this example, the system determined 80% of the data were already cached on DC2 and DC3, 20% of the data should be requested and downloaded;

(2) Then, according to the "near data computing" heuristic rule, the processing tasks are assigned preferentially to the data centers who have already cached data. So the GN processing tasks will be scheduled on DC2 and DC3;

(3) Next, MDCPS selects a suitable data center to download, and assigns the GN processing tasks of the 20% non-cached data by our scheduling algorithm based on Min-Min. In this example, MDCPS firstly determined

that only DC2, DC3, DC5 could provide the services of download for non-cached data based on the data service system. Then, MDCPS calculated the *ECT* of DC2, DC3 and DC5 to process the 20% non-cached data. The methods of time estimation are as previously described. Finally, a data center with the minimum *ECT* should be selected to execute tasks. Here, we assume DC5 was the final selection. The dynamic scheduling process is shown in Figure 6.10;

(4) Finally, L3GN order is split into three sub-orders L3GN1, L3GN2 and L3GN3, and the processing task of GN will be scheduled to DC2, DC3, and DC5.

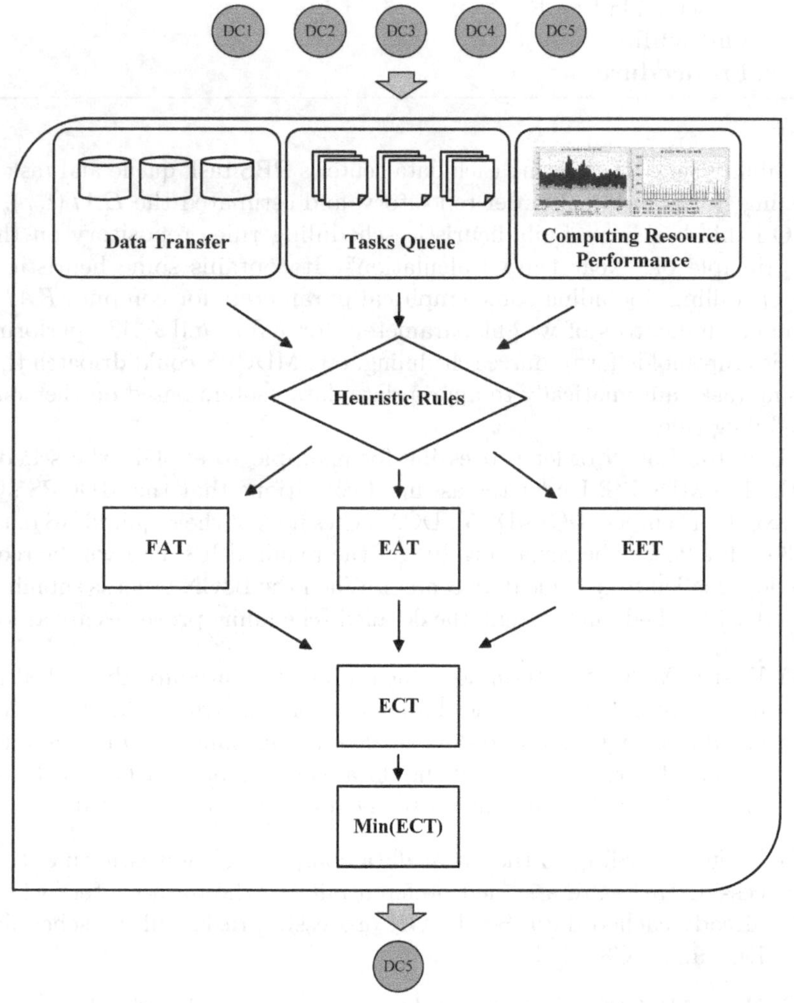

FIGURE 6.10: The procedure of workflow dynamic scheduling in MDCPS.

6.4.2.3 Workflow fault-tolerance

To ensure the reliability of the workflow, the fault-tolerance policies should consider the following aspects:

- Fault-tolerant based on retry: in the construction phase in the workflow, if workflow can't be built correctly caused by the resource being temporarily unavailable, the system will retry build after a certain time interval. The reasons for this kind of fault-tolerance mainly include: unsuccessful webservice call caused by network congestion, insufficient disk space while waiting for cleanup, excessive tasks in the data center, data center overload and data resource in the ready state;

- Fault-tolerance based on checkpoint recovery: In order to ensure a fast reboot of the complex process flow, and to avoid duplication of data transmission and computing, each sub-workflow's condition will be monitored, and the parameters and conditions information will be updated to the task orders libraries at the completion of the sub-task. As soon as the workflow errors occur in the implementation phase, the system will automatically check the recently completed state and rebuild the unfinished task, then resume operation;

- Timeout-based exit strategy: The system will set a threshold, the longest wait time of workflow and PBS job, to avoid long-term occupation of computing resources due to the abnormal operation of workflow; if it goes over the threshold, the workflow or algorithm will stop automatically, and the resources occupied will be recovered.

6.5 Experiments

In order to verify the validity of distributed remote sensing data management and complex workflow management in MDCPS, we conducted the following experiments of performance comparison on several specific data productions.

In this study, we constructed our experimental MDCPS environment with four distributed datacenters: CRESDA, NSOAS, CNIC and RADI. RADI is the MDC, consisting of two compute nodes: one is for workflow management and the other one is for post-processing. The former which is a blade server is configured with 8 cores Intel(R) Xeon(R) E5-2603 (1.80GHz) and 32 GB memory. The latter is configured with 24 cores Intel(R) Xeon(R) E5645 CPU (2.40GHz) and 62 GB memory. All TEPS of CRESDA, NSOAS and CNIC are configured with 16 cores Intel(R) Xeon(R) E5-2640 CPU (2.00GHz) and

32 GB memory. The operating system is CentOS 6.5 and the Java version is 1.8.05.

The groups of experiments produced two kinds of RS product, 1KM Normalized Vegetation Index (NDVI) product and 1KM Net Primary Productive force (NPP) product, during the 180th-185th day of 2014, Zhangye, Gansu Province, China (36° E-43° E, 95° N-103° N) . The NDVI experiment requires 11.5GB of MODIS and FY3 RS data, and the NPP needs 168.7 GB of MODIS, MST2 and FY3 RS data. The results of 1KM NDVI and 1KM NPP products produced by MDCPS are shown as Figure 6.11 and Figure 6.12.

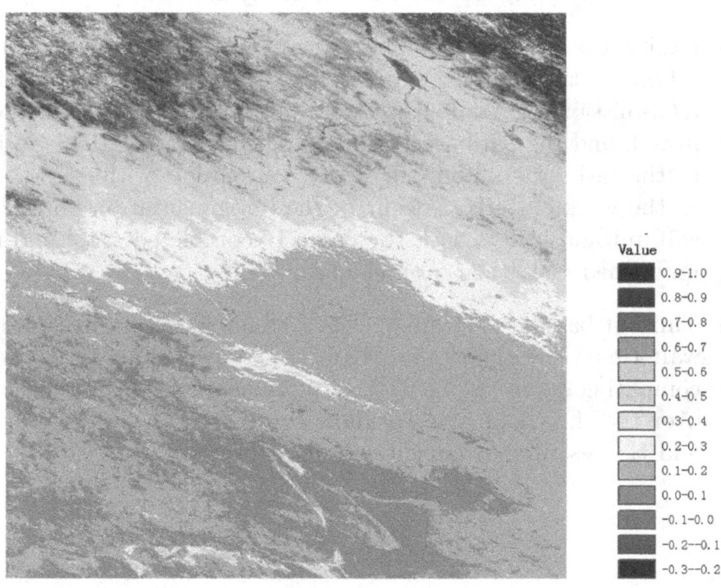

FIGURE 6.11: 1KM NDVI product produced by MDCPS.

6.5.1 Related experiments on dynamic data management

MDCPS realized the dynamic management of the raw data, intermediate products, and the final products, and reduced the large-scale data transmission and repetition. In order to verify the validity of reducing data transmission and data processing in MDCPS, we adopted the following comparative experiments to validate the effect of multi-level RS data cache management. In MDCPS, the production of NDVI and NPP contains seven processing stages: DP, GN, RN, DS, SPU, CPP and CPR, of which, DP and SPU are two stages of data transmission, GN, RN, CPP and DS are four stages of RS algorithmic processing. Generally, distributed massive RS data processing is a data-intensive

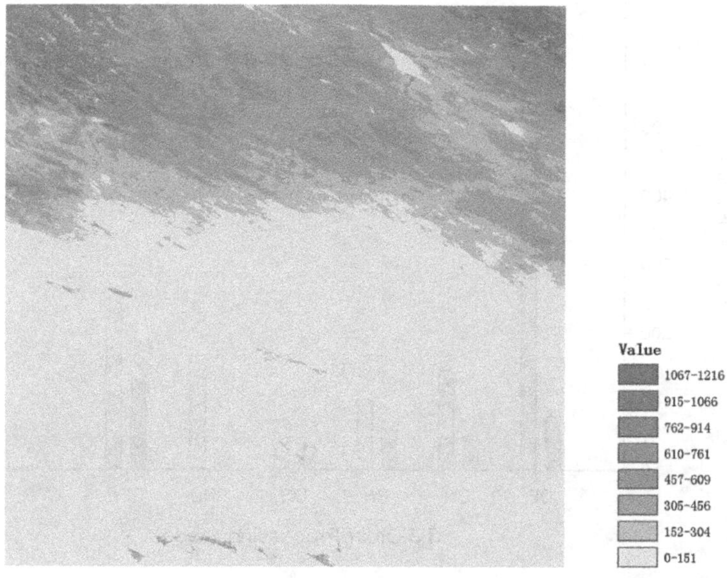

FIGURE 6.12: 1KM NPP product produced by MDCPS.

computing, data transmission and RS algorithmic processing are very time-consuming.

Firstly, we carried out a normal production of NDVI and NPP in MDCPS. This normal production wouldn't reuse any cached data and product, and it is a typical scenario in other distributed data processing systems over WAN. The time-consumption statistics for each stage in normal production of NDVI and NPP is shown in Figure 6.13. It is easy to find out that the runtime of data transmission occupies a larger proportion, the proportion of NDVI production process is about 29.2% and the NPP production process is up to 56.3%. The runtime of algorithm processing also accounted for a considerable proportion, the proportion of NDVI production process is about 69.5% and the NPP production process is 43.3%.

Secondly, we carried out other three typical scenarios based on the same production experiment in the MDCPS system. These productions reused different levels of data cache in MDCPS system, including: raw data cached (80% hit), intermediate products cached (80% hit) and final products cached (100% hit). Compared to the normal production, the processed statistical results are shown in Figure 6.14 and 6.15. By comparison, we can see that the dynamic management of the final products could maximize production efficiency with little time to feedback final products. By comparing the time-consumption of cache raw data and cache intermediate products production scenarios, we could

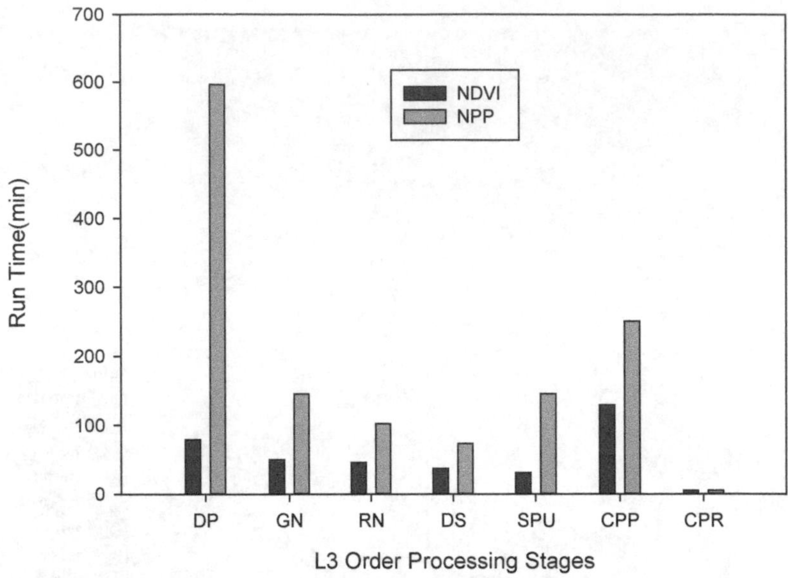

FIGURE 6.13: Runtime of each stage of NDVI and NPP normal production.

find that the raw data cache can only reduce the time-consumption preparation in the DP stage, because it can avoid the repeated request of the same data from the satellite data center. But the data transfer (mainly at the stage of SPU) and algorithmic processing are still time-consuming. The intermediate products caching is better than the raw data caching to enhance the efficiency of production. Its effect is obvious both in the stage of data transmission and algorithm processing. The total time-consumption of intermediate products cached (80%hit) is generally a quarter of that in normal process (NDVI process was 32% and NPP process was about 18%).

According to the above two experiments, we can conclude that multi-level data caching strategies in MDCPS can reduce data transmission and repetition in varying degrees. It can significantly improve the efficiency of production in the multi-datacenter environment.

To test the processing extensions performance of MDCPS for different amounts of data, we tested time-consumption by processing 11.5 GB, 52 GB, 168.7 GB, and 209.2 GB input data for NDVI production. According to the results shown in Figure 6.16, as the amount of process data increases, it shows that time-consumption will increase non-linearly. It could lead to the conclusion that MDCPS has a certain degree capacity for massive data processing.

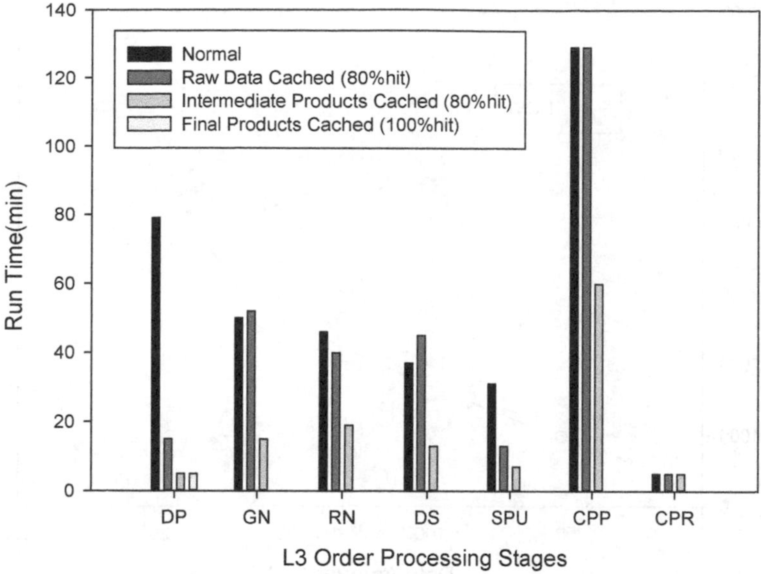

FIGURE 6.14: Runtime of each stage of NDVI with four different scenarios.

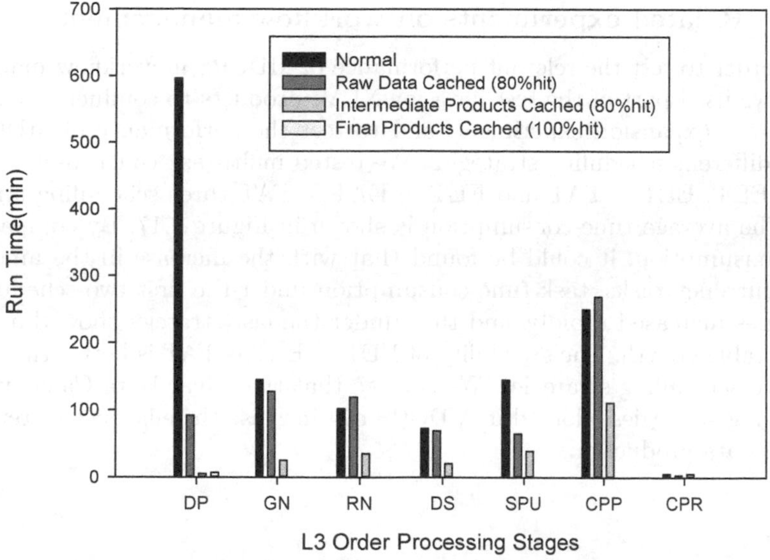

FIGURE 6.15: Runtime of each stage of NPP with four different scenarios.

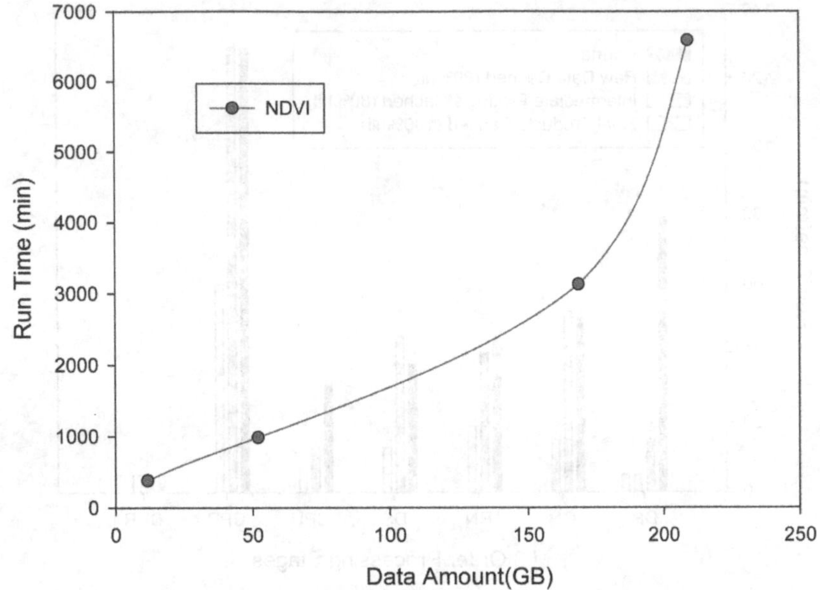

FIGURE 6.16: Runtime of NDVI with the increase of data volume.

6.5.2 Related experiments on workflow management

In order to test the relevant performance of MDCPS in workflow management, we used part of the test data for NDVI products to conduct a related concurrent expansion experiment, and to test the performance of MDCPS under different scheduling strategies. We tested multi-task concurrent scenes under EET, EET + FAT and EET + EAT + FAT three scheduling strategies. The average time-consumption is shown in Figure 6.17. By comparing time-consumption, it could be found that with the increase in the amount of concurrency tasks, task time-consumption under the first two scheduling strategies increased rapidly, and that under the last strategy showed a relatively stable growth. The scalability of EET + EAT + FAT is better than the first two scheduling strategies. We can see that the "Near Data Calculation" scheduling strategies adopted in MDCPS can increase the efficiency of remote sensing data production.

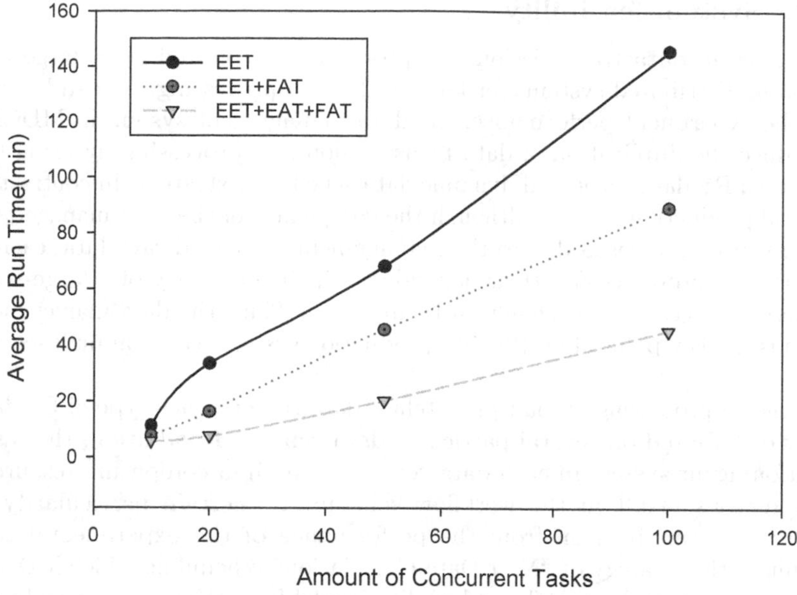

FIGURE 6.17: RunTime of different workflow scheduling scenarios of NDVI production.

6.6 Discussion

6.6.1 System architecture

MDCPS adopts a centralized system architecture to achieve the management of distributed multi-source RS data and production tasks. This centralized system architecture is easy to implement. It not only makes good use of the existing centralized grid middleware such as Globus Grid Security Infrastructure (GSI), GridFTP, Ganglia and Gfarm to build, but also has little influence on the current architecture of data centers; just adding a task execution agent system in the satellite data centers could meet the needs of large-scale production. In addition, the centralized management of workflow can be efficient in global task decomposition and scheduling, and it will significantly improve production efficiency. To avoid the single point of failure in a centralized system, MDCPS could resolve this problem by using the redundant backup of the metadata in back-end databases and distributed replicas in Gfarm.

6.6.2 System feasibility

In general, data transmission and processing accounted for a large proportion in distributed systems for large-scale data processing. As can be seen from the experiment performance, the data management system in MDCPS can reduce the duplication of data transmission and processing by using the distributed RS data cache and dynamic data scheduling strategy. In contrast to a general production system, although the complexity of the data management system will be increased due to the management of cached raw data, cached intermediate products and their dependencies, the efficiency of a large-scale production process can be significantly improved. Thus, the data management system is the key point of an RS data production system based on multi-center architecture.

In the co-processing of multiple satellite data centers, each type of RS data is only distributed on several particular data centers. In addition, the tasks execution agent system in each data center has limited computing resources. These reasons result in the workflow scheduling's certain particularity in MDCPS. As can be seen from the performance of the experimental task scheduling, the strategy of "Near Data Calculation" scheduling which MDCPS adopted, optimizes the workflow scheduling model from the perspective of data transmission, task queue status and the performance of computing resources. The model of multi-objective optimization scheduling is applicable to data-intensive computing in multi-datacenter co-processing.

6.6.3 System scalability

MDCPS has achieved the unified management of data resources, computing resources and algorithm resources, which plays an important role in expanding its ability. When the system wants to add new resources, it only needs to deploy the TEPS in the data center and register its metadata information to the master data center. It can easily integrate new multi-source remote sensing data, processing algorithms and computing resources. And its extension tools, such as automated deployment scripts, will help users expand the system on a cluster or cloud computing platform quickly.

6.7 Conclusions and Future Work

Constructing a remote sensing data processing system based on multiple satellite data centers infrastructure is an effective solution to the problem of massive multi-source remote sensing data processing and analysis. And it is important to support the large scale and global remote sensing application projects. In this systematic project, data management and workflow management are the key issues to build a reliable and efficient distributed processing system.

This chapter summarized the current status of distributed remote sensing data processing across multiple satellite data centers and analyzed the reasons for low efficiency in co-processing from the perspective of data management. In order to solve the problems of massive data migration, we presented a distributed caching strategy of the raw data, intermediate products and final products. Combined with the Gfarm distributed file system, we implemented a distributed data management system in MDCPS. Aiming at the problems of distributed process task management, we completed the decomposition of complex processing tasks by processing repositories. With the help of multi-level orders task repositories and the Kepler workflow template, we achieved automated workflow construction. In addition, we designed a two-level distributed scheduling framework for dispatching processing tasks. The NDVI and NPP production experiments showed that the distributed remote sensing data caching and the scheduling strategies of "Near Data Calculation" could significantly improve overall efficiency.

In the future, more work will be done to better meet the massive remote sensing data production needs based on MDCPS, including developing the user-defined knowledge repositories of remote sensing data processing, providing a service for users to define their own processing workflow based on Kepler, optimizing the knowledge base of RS data production and the heuristic scheduling rules by using the intelligent mobile agent technique, and improving the performance of data distribution strategies to optimize the infrastructure and services of MDCPS.

Chapter 7

Remote Sensing Product Production in an OpenStack-Based Cloud Computing Environment

7.1 Introduction

In recent times, there has been a sharp increase in the number of active and passive remote sensors being sent to space. Those sensors generally have characteristics such as hyper spectral, high spatial resolution, and high time resolution, thus, resulting in a significant increase in the volume, variety, velocity and veracity of data. Due to the richness of the data collected, their applications have also expanded. There are, however, limitations in existing remote sensing data management, processing, production and service pattern systems to adequately deal with the increased demands.

In the data processing system, for example, the data processing capability has not kept pace with the amount of data typically received and needing to be processed. A similar observation is reported by Quick and Choo [184], who remarked that "Existing forensic software solutions have evolved from the first generation of tools and are now beginning to address scalability issues. However, a gap remains in relation to analysis of large and disparate datasets. Every year the volume of data is increasing faster than the capability 'of' processors and forensic tools can manage". For example, the amount of data received from a GF-2 satellite PSM1 sensor is approximately 1.5 TB per day, and the correction of an image (7000*7000*4 pixels) includes billions of floating point operations that require several minutes or even up to an hour to complete using a single workstation. It is clear that the processing time is not appropriate for various real-world applications.

In a production system, existing remote sensing product services primarily support moderate-resolution imaging spectroradiometer (MODIS) [185] and Landsat [186, 187] production, and these services are not capable of providing users with a variety of remote sensing data sources for selection. If users need a multi-source remote sensing data product or another specific product, then they need to search for and download related data and use their own workstation. However, if the user's computing capability and/or knowledge on remote sensing (as well as their computer skills) are limited, then it would be challenging for such a user to obtain the remote sensing products they need [188, 189]. Hence, this motivates the need for a well-designed platform with new data processing system architectures and service patterns to provide products and services for both skilled and unskilled researchers.

Cloud computing has been identified as a potential solution to address some of the big data challenges in remote sensing [190, 191, 192, 193] and big data computing [194, 78, 57, 195], such as allowing massive remote sensing data storage and complex data processing, providing on-demand services [196, 197, 198], and improving the timeliness of remote sensing information service delivery. For example, Lv, Hu, Zhong, Wu, Li and Zhao [199] demonstrated the feasibility of using MapReduce and parallel K-means clustering for remote sensing image storage and processing. Also using MapReduce, Almeer [200] built an experimental, high-performance cloud computing system in the Environmental Studies Center at the University of Qatar. Lin, Chung, Wang, Ku and Chou [201] proposed and implemented a framework desigend to store and process massive remote sensing images on a cloud computing platform. Similarly, Wang, Wang, Chen and Ni [193] compared the use of rapid processing methods and strategies for remote sensing images, using cloud computing and other computing paradigms. However, the focus of these studies is only on the storage and computational capabilities using cloud computing, rather than product generation and the information service of remote sensing.

Above all, the existing remote sensing systems are facing the following major issues: (1) data processing capability has not kept pace with the amount of data typically received and needing to be processed; (2) product services are not capable of providing users with a variety of remote sensing data sources for selection, and a well-designed platform is urgently needed to provide products and services for both skilled and unskilled researchers; (3) the current cloud-based remote sensing computing is less focused on product generation and information service. Therefore, in order to tackle these complex challenges, in this study, we present a product generation programme using multisource remote sensing data, across distributed data centers in a cloud environment. This allows us to achieve massive data storage, high-performance computing, virtualization, elastic expansion, on-demand services and other cloud-inherent characteristics. We also provide an easy-to-use multi-source remote sensing data processing and production platform [202]. Finally, we demonstrate the utility of our approach using data processing and production generation experiments.

The remainder of this chapter is organized as follows. In the next section, we provide an overview of the background and related work. Section 7.3 introduces the proposed cloud-based programme framework, system architectures, business logic and service patterns. Section 7.4 describes the experiments and study cases. Finally, in Section 7.5, we provide a summary and conclude the chapter.

7.2 Background and Related Work

This section briefly reviews remote sensing products and production system architectures, and related work.

7.2.1 Remote sensing products

Remote sensing products can be broadly categorized into fine processing products, inversion index products, and thematic products.

7.2.1.1 Fine processing products

Fine processing products mainly include geometric normalization products, radiometric normalization products, mosaic products, and fusion products.

- Geometric normalization products refer to geometric-registered image collection, where using geometric precision corrections, the images are turned into space seamless remote sensing products [203].
- Radiometric normalization products are quantitative remote sensing products (essentially, products obtained after radiometric cross-calibration, long time series radiometric normalization, atmospheric correction, etc.) [204].
- Mosaic products can be explained simply as stitching two or more orthorectified satellite images with an overlapping area if the images from the satellite do not include atmospheric effects. To create a mosaic of two or more optical satellite remote sensing images, we geometrically correct the raw optical remote sensing dataset to a known map coordinates system (e.g., geographic coordinates system or projected coordinates system) as well as preprocessing the atmospheric corrections (e.g., image-based model, empirical line model and atmospheric condition model). However, it is important to consider apparent seasonal changes in order to stitch mosaic images obtained from different seasons, due to the difficulties in acquiring high-resolution optical images in the same season for a number of reasons such as adverse weather conditions [205].
- Fusion products, one of the most commonly used remote sensing data, integrate information acquired with different spatial and spectral resolutions from sensors mounted on satellites, aircraft and ground platforms, and they contain more detailed information than each of the sources [19]. Fusion techniques are useful for a variety of applications, ranging from object detection, recognition, identification and classification, to object tracking, change detection, decision making, etc. [206].

7.2.1.2 Inversion index products

Inversion index products generally refer to various inversion products of geophysical parameters, which reflect variation in characteristics of the land, sea and weather, such as the Normalized Difference Vegetation Index (NDVI) [207], Normalized Difference Water Index (NDWI) [208], Normalized Difference Drought Index (NDDI)[209], Normalized Difference Build-up Index (NDBI)[210], and Normalized Difference Snow Index (NDSI)[211].

7.2.1.3 Thematic products

Thematic products are application-oriented products or maps, such as thematic land-use and mineral thematic maps. Thematic products are generally obtained through remote sensing image interpretation and a remote sensing inversion model, as well as expert knowledge.

In general, the above mentioned remote sensing products have upper and lower hierarchical relationships. That is to say, if we wish to obtain a remote sensing thematic product, the fine processing or inversion index products may be generated first. According to this hierarchical relationship, we build a knowledge base of remote sensing products and their corresponding production parameters, which guide the products generation.

7.2.2 Remote sensing production system

The remote sensing production system architecture can be broadly categorized into (1) personal computers (PCs) or a single workstation that acts as the remote sensing processing system, and (2) a high-performance cluster-based remote sensing processing system. The stand-alone processing system uses the computational resources of a single computer to independently perform remote sensing data processing and a production process via human-computer interaction. As computing technologies develop over the years, the stand-alone processing system has evolved from big- and medium-sized computers specializing in remote sensing data processing to super-minicomputers specializing in remote sensing data processing to PC-based universal remote sensing data processing systems with multi-core processors to the current hyper-threading Graphics Processing Units (GPUs) [212, 213] universal remote sensing data processing systems [214].

The cluster-based remote sensing processing system generally consists of a number of identical PCs connected to multiple networks, tied together with channel bonding software. Thus, the networks act like one network running at many times the speed [215, 216]. Notable examples include "Pixel Factory" (France's massive remote sensing data processing platform), the Grid Processing on Demand (G-POD) European Space Agency project [217], Global Earth Observation System of Systems (GEOSS) [218], and the Parallel Image Processing System (PIPS) of the Chinese Academy of Sciences Institute of Remote Sensing and Digital Earth [219, 220].

With a significant increase in remote sensing data, methods for remote sensing data processing and product generation are often limited by the sheer volume of data and computational demands that far exceed the capability of single workstations. Cluster-based High-performance computing (HPC) had been used to rapidly analyze very large data sets (> 10 terabytes can be rapidly analyzed); thus, in this study, we build an HPC cluster with Open MPI in the cloud environment. This allows us to process global data sets, detect environmental change, and generate remote sensing products.

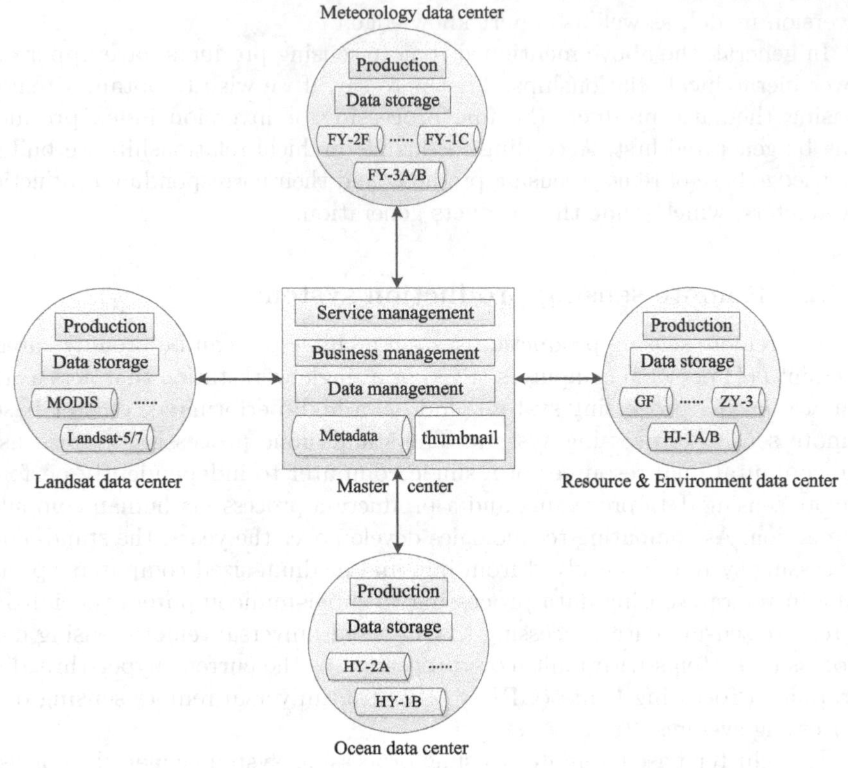

FIGURE 7.1: Cloud RS framework.

7.3 Cloud-Based Remote Sensing Production System

7.3.1 Program framework

Our proposed Cloud-based Remote Sensing (Cloud RS) programme adopts a master and multiple slaves' architecture (see Figure 7.1). The master center is mainly responsible for the production order receiving and parsing, task and data scheduling, results feedback, and so on. The slave centers are the distributed remote sensing data centers, which store one or more types of remote sensing data. These slave centers are also mainly responsible for production task execution. In addition, the master center ingests metadata and thumbnails of the remote sensing data from each slave center, and the service portal, including data service, production service and cloud storage service, distributed on the master center.

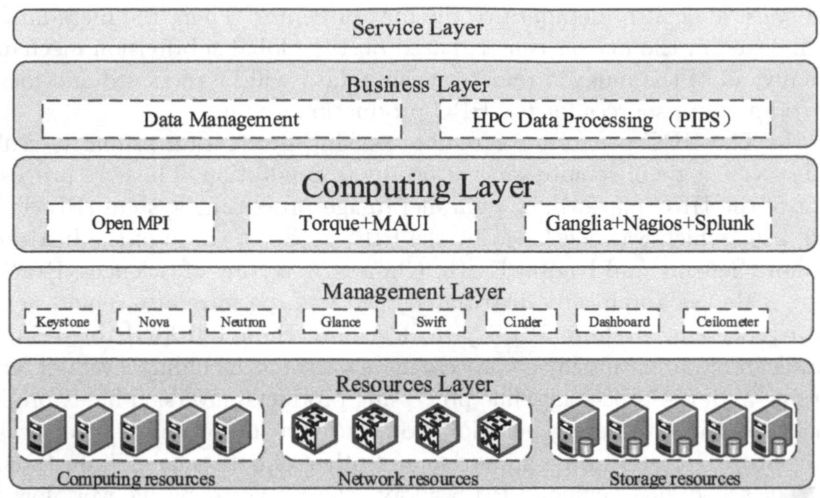

FIGURE 7.2: Cloud RS system architecture.

7.3.2 System architecture

The architecture of the remote sensing production system in the cloud environment consists of five layers, namely: resources, management, computing, business and service (see Figure 7.2).

- The resource layer builds a large number of computing resources, network resources and storage resources connected by networks into virtualized resource pools. Virtual resources such as virtual CPU, virtual memory, virtual disk, virtual object storage space and virtual network that can be uniformly managed within the remote sensing cloud service platform are formed [221, 222].
- The management layer mainly adopts the OpenStack [223] computing framework and uses its core components to manage virtual resources.
- The computing layer mainly provides a virtual cluster computing environment, including service such as massive remote sensing data storage, cluster computing, cluster scheduling and computing environment monitoring.
- The business layer mainly includes two parts, namely: (a) remote sensing data management across distributed data centers and (b) an HPC platform for remote sensing production.

(a) Remote sensing data management mainly includes multi-source data ingesting, metadata index and data retrieval. Data ingesting from distributed data centers is mainly based on a crawler, which will launch itself at regular intervals, and push metadata to the master center. Then, the metadata will be indexed in the master center, based on the global subdivision mechanism. Finally, all of the indexed remote sensing data will be retrieved and located, providing data sources for the HPC production system.

(b) The HPC platform for remote sensing production primarily realizes the service logic of remote senisng products generation. The core processing unit of the HPC platform is a parallel image processing system (PIPS) [224, 225, 226], which was developed by the PIPS research group of the Institute of Remote Sensing and Digital Earth, Chinese Academy of Sciences. PIPS is a large-scale, geographically distributed, and high-performance remote sensing data processing system. PIPS provides more than 100 kinds of serial and parallel remote sensing image processing algorithms, including level 0-2 remote sensing data pre-processing, fine processing products, inversion index products and thematic products generation, so as to provide production services for agriculture, forestry, mining, marine and other remote sensing industries.

The scheduling engine of PIPS adopts the Kepler scientific workflow [202, 227], and each production workflow mainly includes three parts: data preparation, workflow organization and production task execution.

Data preparation denotes the input data selection of each production order. In general, the production needed data is the standard remote sensing data, which is geometric and radiometric normalized original data, and their metadata are stored in the standard database. It's important to note that the original data is level 1B or level 2 remote sensing data, and they all have been, after inter-detector equalizations (sometimes referred to as radiometric correction) and systematic geometry correction. But as for some higher-level inversion index products, their required data may be some lower-level inversion products, and the data selection strategy may be very complex in this situation. For example, the required data sources of Net Primary Productivity (NPP) are Photosynthetic Active Radiation (PAR) and Leaf Area Index (LAI), but PAR and LAI still need some vegetation indexs (VIs), and the VIs require standard remote sensing data as input data sources; this is very complex.

Workflow organization is essentially to determine the execution sequence of each process unit. In Kepler, the process unit is viewed as an Actor, and it was linked together by Relation and Link. Link determines the input and output of each Actor, and Relation shows the upper and lower hierarchical relationship of each Actor. For example, Figure 7.3 shows the Kepler model corresponding to the NDVI workflow. For the clarity of the chapter, we omitted several steps from the original workflow. There are two inputs of the process unit NDVI calculation module, red and near-infrared bands, and output is a product. The upper module of NDVI is a GN module, and the lower module is product register. With the same procedure other vegetation indexes or image processing workflows can be implemented. However, there are so many kinds of remote sensing products that we cannot list all of their Kepler workflows, or some

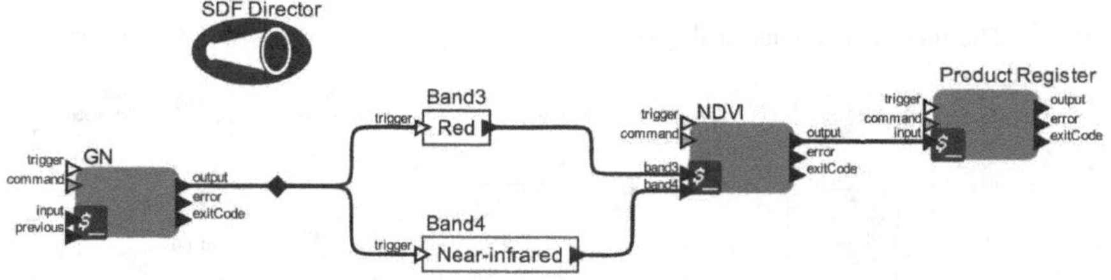

FIGURE 7.3: The Kepler workflow of NDVI.

users may want to define their production workflows. Therefore, we should provide a workflow auto-building capacity; then the knowledge database is built. And the knowledge base will be detailed in the next sub section.

After data preparation and workflow organization, the production tasks enter the implementation phase. In general, the production task runs only on one data providing center, and the data scheduling adopts the minimum data transferring strategy if one center cannot satisfy the data demand of the production task. As for the computing nodes selection on each data center, the total number is determined by the production task priority, or user determined, and which nodes assigned are determined by the scheduling policy of TORQUE and MAUI, as well as the resources monitoring results of Ganglia. In addition, in order to improve the robustness of the whole production system, a self-management function is added.

Furthermore, cloud services [228, 229] include user registration and authentication services, user charging services, remote sensing data and product services, cloud storage services, etc.

7.3.3 Knowledge base and inference rules

Our knowledge base mainly includes three parts: (1) the upper and lower hierarchical relationship database (2) the input/output database of every kind of remote sensing product and (3) inference rules for workflow organization.

7.3.3.1 The upper and lower hierarchical relationship database

As mentioned before, remote sensing products mainly include three classifications, fine processing, inversion index and thematic products. As for each level, there are still some sub level products (Figure 7.4). In order to better use the hierarchical relationship of remote sensing products to organize Kepler production workflow, we encoded each product, and the coding rules are as follows.

- The first layer coding includes two binary codes, '0' and '1', and '00', '01', '10', '11' denote the 'original image', 'fine processing products', 'inversion index products' and 'thematic products' separately.

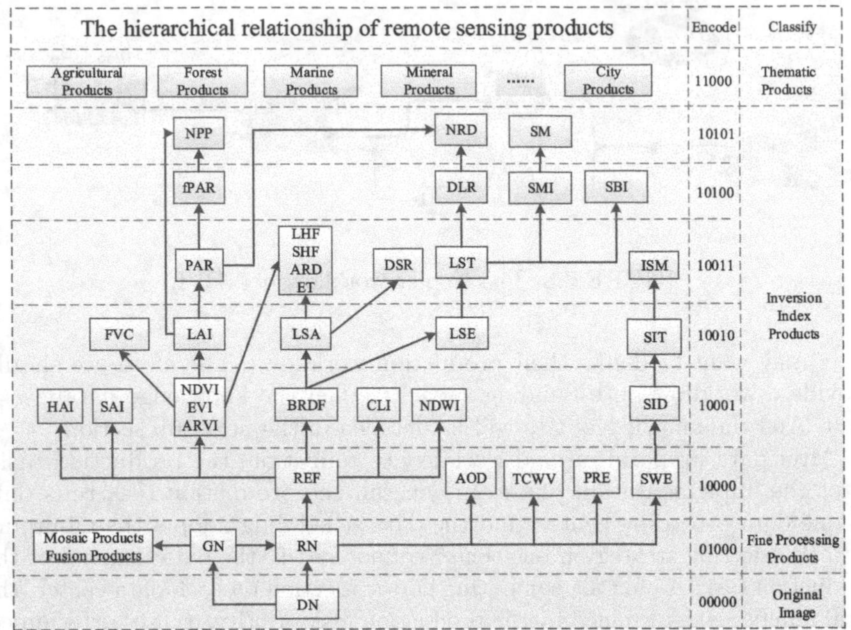

FIGURE 7.4: The upper and lower hierarchical relationship of remote sensing products.

- The second layer coding mainly aims for the sub level products of each type. In order to as much as possible include all of the sub level products, we chose three binary codes, and the concrete coding results are as shown in the right part of Figure 7.4.
- The third layer coding is for the product in each sub level; i.e., this layer coding is an ID of each product. This layer coding uses four binary codes.

Above all, as for each remote sensing product, the whole coding will include 9 binary codes, and the final results are as shown in Table 7.1.

7.3.3.2 Input/output database of every kind of remote sensing product

In addition to the upper and lower hierarchical relationship, we need to consider the input and output of each remote sensing product, so as to construct the production workflow. As mentioned above, remote sensing production needs one or more types of standard remote sensing data, and some inversion index products. Therefore, we build the input/output database for every kind of remote sensing product. Taking NPP as an example, the input/output database is shown in Figure 7.5.

TABLE 7.1: The coding results of remote sensing products (part).

product name	short name	coding
Digital Number	DN	000000000
Radiometric Normalized	RN	010000000
Geometric Normalized	GN	010000001
Mosaic Products	Mosaic Products	010000010
Fusion Products	Fusion Products	010000011
Surface Reflectance	REF	100000000
Silicide Anomaly Index	SAI	100010001
Normalized Difference Vegetation Index	NDVI	100010010
Enhanced Vegetation Index	EVI	100010011
Atmospherically Resistant Vegetation Index	ARVI	100010100
Bidirectional Reflectance Distribution Function	BRDF	100010101
Normalized Difference Water Index	NDWI	100010111
Sea Ice Distribution	SID	100011000
Vegetation Fractional Coverage	FVC	100100000
Leaf Area Index	LAI	100100001
Land Surface Albedo	LSA	100100010
Photosynthetically Active Radiation	PAR	100110000
Evapotranspire	ET	100110001
Aerodynamic Roughness Length	ARD	100110010
Sensible Heat Flux	SHF	100110011
Downward Shortwave Radiation	DSR	100110101
Land Surface Temperature	LST	100110110
Ice Snow Mass Change	ISM	100110111
Fraction Of Photosynthetically Active Radiation	fPAR	101000000
Soil Moisture Index	SMI	101000010
Soil Brightness Index	SBI	101000011
Net Primary Productivity	NPP	101010000

7.3.3.3 Inference rules for production demand data selection

In order to prepare data sources for production, we established a set of inference rules, and it is as shown in Figure 7.6.

As shown in Figure 7.6, each production order needs to parse inputParametersData, inputParametersProducts and auxiliaryData, three types of parameters, and they correspond to the standard data, products and auxiliary data (Auxiliary data is only ready for some particular products). After parsing and reasoning, each type of data name and location will be returned, preparing for production scheduling.

7.3.3.4 Inference rules for workflow organization

Production algorithms all have a one-to-one correspondence relationship with remote sensing products. Hence, the organization of remote sensing production workflow is based on the upper and lower hierarchical relationship database and input/output database. We also established inference rules in order to provide guidance for Kepler workflow self-organization (Figure 7.7).

```
<inputParametersData>
    <data>
        <inputdatatype>1</inputdatatype>
        <satellite>TERRA/AQUA </satellite >
        <sensor>MODIS</sensor>
        <productweight>0.9</productweight>
    </data>
</inputParametersData>
<inputParametersProducts>
    <product>
        <productTag>0</productTag>
        <productID>LAI</productID>
    </product>
    <product>
        <productTag>1</productTag>
        <productID>PAR</productID>
    </product>
    <product>
        <productTag>2</productTag>
        <productID>FPAR</productID>
    </product>
</inputParametersProducts>
```

FIGURE 7.5: The input/output database for remote sensing products (NPP as an example).

7.3.4 Business logic

The business logic of a remote sensing production system mainly includes order submit, order analysis, completeness analysis, data preparation, products generation, products management and other related processes (see Figure 7.8).

An order analysis business module analyzes the feasibility of user-submitted product orders, which depends upon whether there is potentially needed original remote sensing data in the database, whether there are potentially needed products in the products database, and whether there are potentially needed

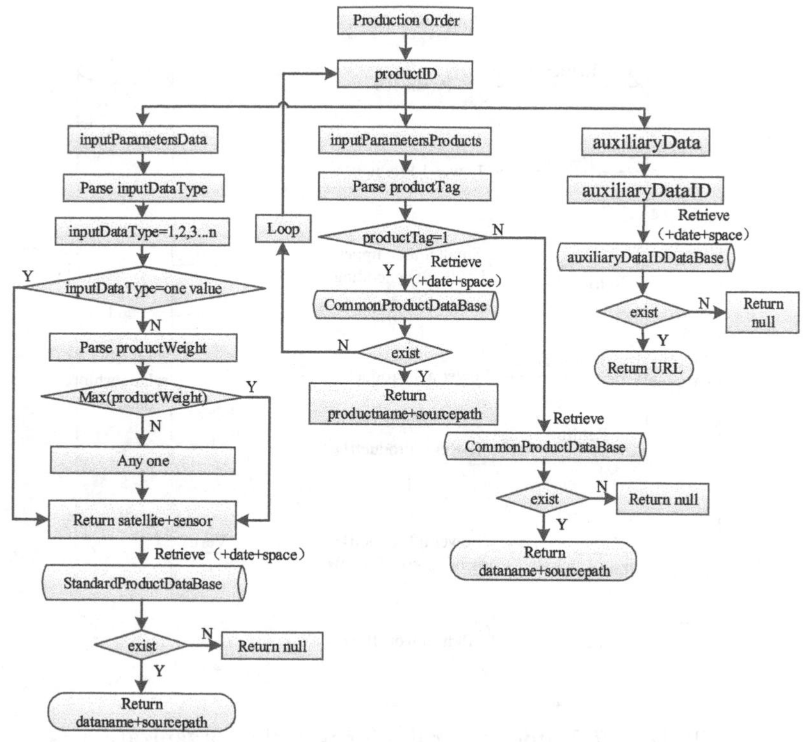

FIGURE 7.6: Inference rules for production demand data selection.

workflows in the workflows database. If the result of the order analysis is successful, then the remote sensing data preparation will arrive.

Remote sensing data preparation is aimed at determining and preparing the kinds and amounts of potentially needed original data according to the order analysis results. In essence, the original data refers to the preprocessed remote sensing data, which is available after radiometric correction and systematic geometry correction. It should be noted that data preparation is based on a good data management mechanism. In our system, the data management includes the original remote sensing database, database index, distributed storage strategy of multiple copies of remote sensing data, etc.

After data preparation, completeness analysis for the prepared remote sensing data is essential. Completeness analysis includes time range completeness analysis and space range completeness analysis. If the result of the completeness analysis is true, then the prepared remote sensing data will be transferred into the products generation module. But first, the prepared data should be standardized, including radiometric and geometric normalized. After normalization, the product orders will be processed. Remote sensing products

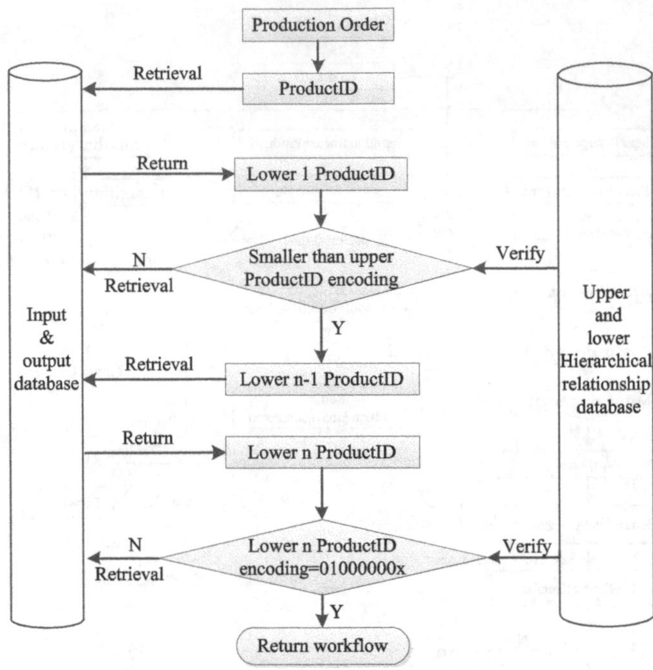

FIGURE 7.7: Inference rules for workflow organization.

generation business module includes workflow and corresponding algorithm selection, computing resource allocation, production task self-management and so on. Workflow and corresponding algorithm selection are determined by the products knowledge base, which has been detailed earlier. The computing resource allocation is mainly based on the workflow complexity index and real-time resource monitoring information. The workflow complexity index is calculated with the input data volume, space and time complexities of the algorithms. Real-time resource monitoring information is obtained by Ganglia cluster-monitoring software. Production task self-management includes running task state monitoring, system log monitoring, fault-tolerance of the running job, etc.

Finally, the products will be checked-in and saved in the products management system. Products check-in refers to writing the metadata of the processed products into the products database for direct user product retrieval, so as to avoid repeated production processes.

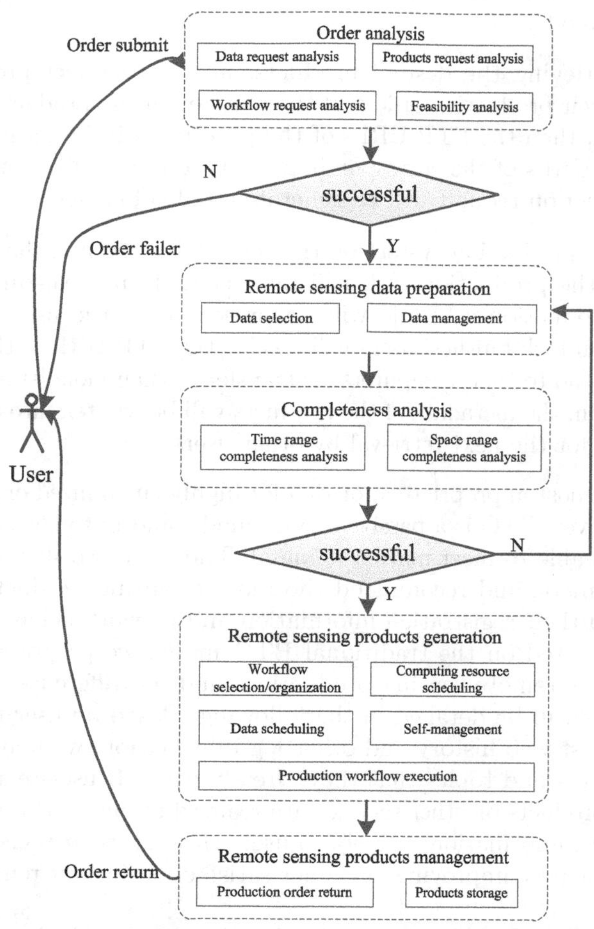

FIGURE 7.8: Business logic of Cloud RS production system.

7.3.5 Active service patterns

In general, the traditional service pattern of remote sensing production is the Build to Order (BTO) mode [230], sometimes referred to as Made to Order (MTO). This is a production approach whereby products are not built until a confirmed order is received, and the steps of its concrete realization are as follow.

- By browsing the service portal, users can learn about remote sensing products information provided on the server side. Then, by the aid of a search engine, users' wanted products will be located, and it is generally accomplished based on the "product type + time range + space range" retrieval mode.

- After retrieving the desired products, users can select products and submit their production orders. If the remote sensing products have been produced, then the FTP URLs of the products will be returned. If not, the FTP URLs of the corresponding original data will be returned, and the production request will be generated and submitted.

- When the production system on the server side receives the production request, the production task will be executed and real-time feedback of the production schedule will be presented to the users. After the production task finishes, users will receive the FTP URL of the products for download to their computers, or transfer to their cloud storage system. In addition, the metadata of the products will be written into the product database for the next retrieval by other users.

BTO is the most appropriate approach for highly customized or low volume products. However, BTO is a passive service mode, and in today's competitive market, it is unable to meet market demand. Therefore, we should design an active service mode and recommend the remote sensing product service to users, based on their registration information and network behavior.

Therefore, based on the traditional BTO mode, we proposed an active service mode for remote sensing production, and its differences from BTO and main ideas will be detailed in the following. Based on user registration information, past web history and other log information, we may determine the types of users and topics that they already want. If user-related remote sensing data, products or other services are realized in our platform, we would push the service information actively to users through their registered email or cellphone, thereby improving customer satisfaction for our remote sensing service.

In our cloud-based remote sensing producttion system, the implementation procedures of the active remote sensing production service can be generated using the Unified Modeling Language (UML), as shown in Figure 7.9.

As can be seen in Figure 7.9, the cloud-based service pattern differs from the BTO mode in two respects:

- Once users' requested products have been prepared, users can choose to save the products into their cloud storage, which is provided by the production system. The advantages of this pattern not only avoid the trouble caused by limited user storage capacity, but also improve the level of remote sensing data sharing.

- The individualization active recommendation information service enhances the utilization efficiency of the new generated products, while the active service pattern can provide personalized services for remote sensing users.

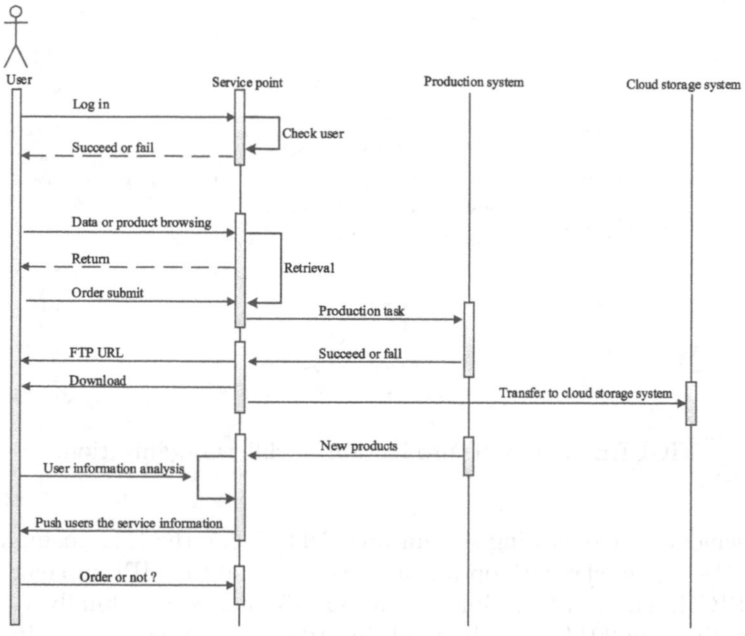

FIGURE 7.9: UML service pattern for the Cloud RS production system.

7.4 Experiment and Case Study

We evaluated the proposed programme using test experiments performed at global, regional and local areas.

7.4.1 Global scale remote sensing production

At the global scale remote sensing production, we chose the higher level inversion product NPP as an example. This is the quantity of carbon dioxide vegetation consumed during photosynthesis excluding the quantity of carbon dioxide the plants release during respiration (metabolizing sugars and starches for energy). In our test experiment, NPP was generated every 5 days in 2014, with 1 kilometer spatial resolution, and its input data sources are mainly MODIS L1B 1KM data from the Landsat remote sensing data center. The production workflow is shown in Figure 7.10.

In order to realize the annual global scale NPP production, the volume of required MODIS L1B 1KM data is about 11 terabytes (TB). Therefore, we provide a virtual multi-core cluster with 10 nodes. Each node is an x-large type of OpenStack instance, with 8 Virtual CPUs (VCPUs) and 16 gigabyte

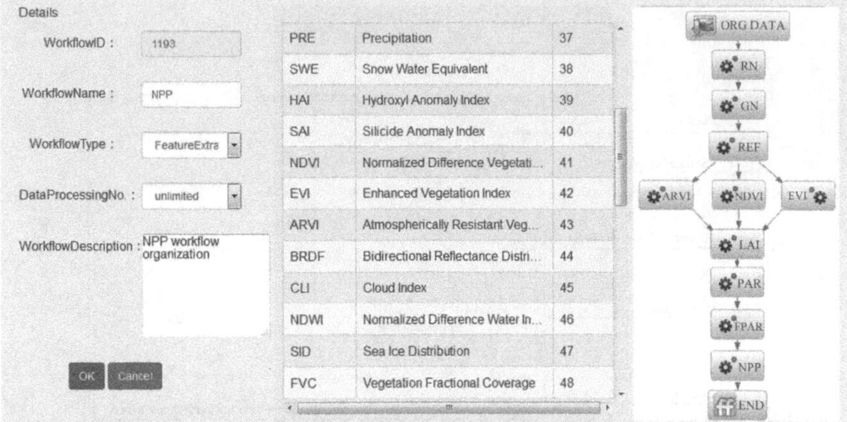

FIGURE 7.10: NPP production workflow organization.

(GB) memory. The operating system was CentOS 6.5, the C++ compiler was a GNU C++ Compiler with optimizing level O3, and the MPI implementation was MPICH. The total runtime was about 135 hours, and finally 74 global NPP products in 2014 were obtained. In order to examine the quality of the generated NPP products, we selected 6 of them in different months, and they are shown in Figure 7.11.

As observed in Figure 7.11, in mid-latitudes, NPP is clearly tied to seasonal change, with productivity peaking in each hemisphere's summer. The Boreal Forests of Canada and Russia, for example, experience high productivity in July and then a slow decline through fall and winter. Year-round, tropical forests in South America, Africa, Southeast Asia, and Indonesia have high productivity, not surprising with the abundant sunlight, warmth, and rainfall. This was well adapted for natural animal growing, and also proved the practicability of our production system.

7.4.2 Regional scale mosaic production

We selected a 7 bands Landsat-TM image as the regional scale mosaic data source, and the spatial scope is north and east China (113°02′E- 123°32′ E, 30°45′N- 42°21′N). The total image number is 28, and the total volume is about 10 GB. The mosaic algorithm adopts a parallel computing solution, and the total runtime with increasing numbers of virtual processors is as shown in Figure 7.12.

As can be seen in Figure 7.12, the total runtime decreases sharply when scaled to less than 32 VCPUs. However, the decrease rate is much slower when scaled from 40 VCPUs to more, and the runtime increases even up to 80 VCPUs. This is probably because the communication time between nodes occupies a great deal of total runtime. The bands 4,3 and 2 false color composite image of the final region scale mosaic product is as shown in Figure 7.13.

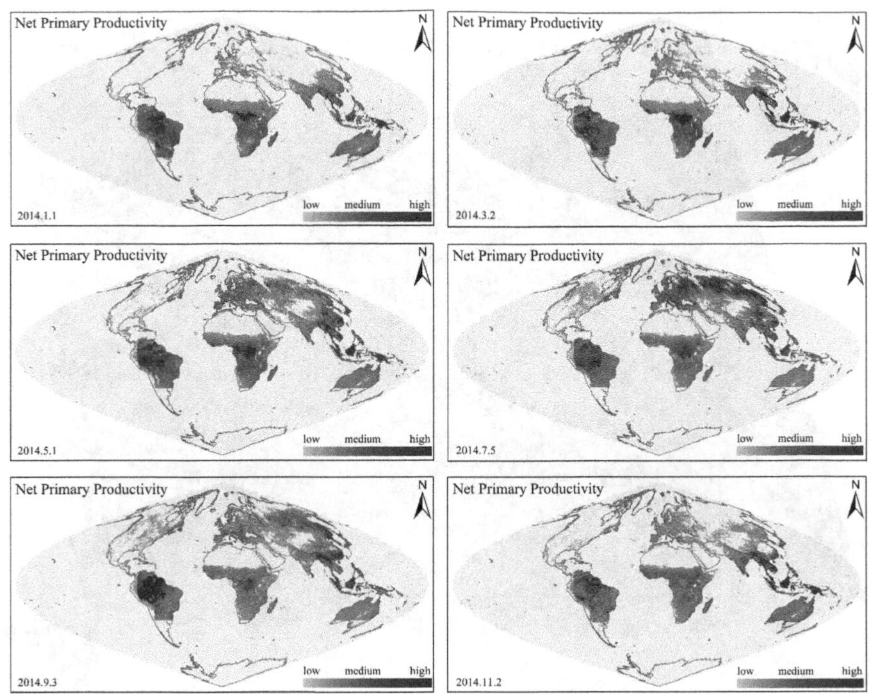

FIGURE 7.11: The global NPP maps in different months.

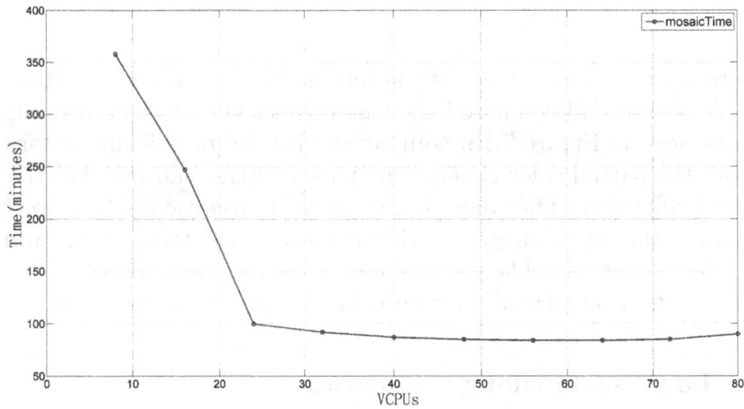

FIGURE 7.12: Total runtime of mosaic production with scaling virtual processors.

FIGURE 7.13: Landsat-TM mosaic product of north and east China(R:band4, G:band3, B:band2).

In order to verify the effect of the mosaic production, we selected a 400 x 400 pixels region, and compared the visual effect before and after mosaicking. As can be seen in Figure 7.13, compared with the mosaicking before image, except for the partial color change, the mosaic image can preserve the border structures efficiently. The color change may be because of the color balance among all of the input images, during the process of mosaicking, and this is difficult but inevitable. Therefore, comprehensively considering the runtime and mosaic effect, our cloud-based production system is powerful.

7.4.3 Local scale change detection

Furthermore, in order to realize the time-series remote sensing production, we provide a data cube technology. But first, we should introduce the remote sensing data cube concept.

7.4.3.1 Remote sensing data cube

After geometric and radiometric normalization, remote sensing data are essentially becoming quantitative image 'tiles', which are two dimensional space seamless grids. Repeated observations of the same area at regular intervals produce a sequence of satellite image 'tiles'. If we collect these 'tiles' in time sequences covering the same areas of ground, it can be visualised as a three dimensional data set with the time axis as the third dimension. This is informally referred to as a 'cube'. The cube can be analysed and used to detect changes in the environment, so as to inform government about the effects of land degradation, flood damage and deforestation.

7.4.3.2 Local scale time-series production

At local scale time-series production, we chose Aibi Lake, which is in the northwest of China, as the study area. From September 18, 2008 to September 18, 2016, the HJ1A/B-CCD remote sensing data of that area, about a total number of 800 original scenes, were ordered from the China Center for Resources Satellite Data and Application (CRESDA). After geometric and radiometric normalization, a subset of each image area (2000 pixels x 2000 pixels x 4) was extracted, which covers Aibi Lake. After weeding out the poor quality data, the remaining subset images were collected in time sequences, and the HJ1A/B-CCD data cube of Aibi Lake was obtained (Figure 7.14)).

The prepared HJ1A/B-CCD data cube contains 421 scene images, each image with 4 bands, 2000 x 2000 x 4 pixels, and the whole cube about 26.4 GB. Using the NDWI production workflow in our system, with 10 nodes virtual computing instances, after about 5 minutes, the final product was as shown in Figure 7.15.

As observed in Figure 7.15, the blue area denotes the frequent water region of Aibi Lake, and the deeper blue color, the higher frequency. Some regions do not have persistent water, and this may be caused by seasonal variation. This is consistent with the natural laws, and further confirmed the practicability of our production system.

7.5 Conclusions

In this study, we briefly reviewed remote sensing products and production system architectures, prior to presenting our cloud-based production system, across distributed data centers. We also described the system architectures, business logic and service patterns. Specifically, the proposed system has a five-layer architecture, which integrates several web and cloud computing technologies. Leveraging the benefits afforded by the use of cloud computing, we are able to support massive remote sensing data storage and processing.

FIGURE 7.14: HJ1A/B-CCD data cube of Aibi Lake (R:band4, G:band3, B:band2).

Each user can use the virtual machine and cloud storage conveniently, thus, reducing information technology resource costs. Moreover, the system adopts the individualization active recommendation information service.

Finally, findings from the test experiments (i.e., global scale production, regional scale mosaic, and local scale time-series analysis) demonstrated the powerful computing capabilities and advantages of our proposed programme. In other words, the proposed cloud-based remote sensing production system can deal with massive remote sensing data and generating different products, as well as on-demand remote sensing computing and information service. Future work includes extending, implementing and evaluating a prototype of the proposed system in a real-world scenario.

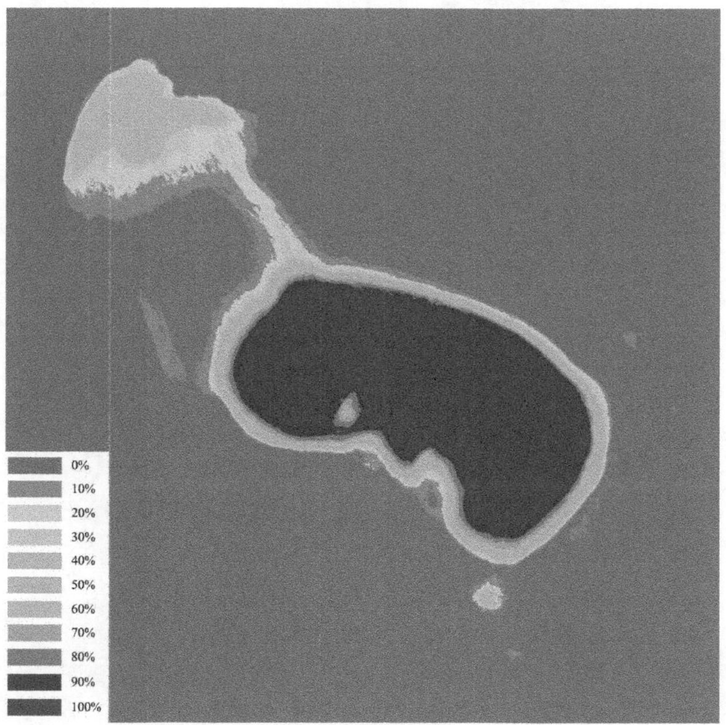

FIGURE 7.15: The 9-year time series results of Aibi Lake (R:band4, G:band3, B:band2).

Chapter 8

Knowledge Discovery and Information Analysis from Remote Sensing Big Data

8.1 Introduction

In recent decades, the remarkable developments in Earth observing(EO) technology provided a significant amount of remote sensing(RS) data openly available[231]. This large observation dataset characterized the information about the earth surface in space, time, and spectral dimensions[232][233]. Apart form these dimensions, these data also contain many geographic features, such as forests, cities, lakes and so on, and these features could help researchers to locate their interested study areas rapidly. Now these multidimensional RS data with features have been widely used for global change detection research such as monitoring deforestation[234] and detecting temporal changes in floods[235]. However, the conventional geographic information system(GIS) tools are inconvenient for scientists to process the multidimensional RS data, because they lack appropriate methods to express multidimensional data models for analysis. And researchers have to do additional data organization work to conduct change detection analysis. For a more convenient analysis, they need a multidimensional data model which could support seamless analysis in space, time, spectral and feature.

Recently, many researchers have proposed using a multidimensional array model to organize the RS raster data[236][237]. Subsequently, they achieved the spatio-temporal aggregations capacity used in spatial on-line analytical processing(SOLAP) systems[238][239], as a data cube. Using this model, researchers can conveniently extract the desired data from the large dataset for analysis, and it reduces the burdens of data preparation for researchers in time-series analysis. However, in addition to extracting data with simple three-dimensional(3D) space-time coordinates, researchers occasionally need to extract data with some geographic features[240][241][242], which are often used to locate or mark the target regions of interest. For example, flood monitoring often needs to process multidimensional RS data which have the characteristics of large covered range, long time series and multi bands[243]. If we built all the analysis data as a whole data cube which has the lakes or river features, we could rapidly find the target study area we need by feature and analyse the multi bands image data to detect the flood situation with the time series. That makes researchers focus on their analysis work without being troubled by the data organization. This study aims to develop the spatial-feature remote sensing data cube(SRSDC), a data cube whose goal is to deliver a spatial-feature-supported, efficient, and scalable multidimensional data analysis system to handle the large-scale RS data. The SRSDC provides spatial feature repositories to store and manage the vector feature data, and a feature translation to transform the spatial feature information to a query operation. To support large-scale data analysis, the SRSDC provides a work-scheduler architecture to process the sliced multidimensional arrays with dask[244].

The remainder of this chapter is organized as follows. Section 8.2 describes some preliminaries and related works. Section 8.3 presents an architectural overview of the SRSDC. Section 8.4 presents the design and implementation of a feature data cube and distributed execution engine in the SRSDC. Section 8.5 uses the long time-series remote sensing production process and analysis as examples to evaluate the performance of a feature data cube and distributed execution engine. Section 8.6 concludes this chapter and describes the future work prospects.

8.2 Preliminaries and Related Work

8.2.1 Knowledge discovery categories

In the following, we discuss four broad categories of applications in geosciences where knowledge discovery methods have been widely used and have aroused impressive attention. For each application, a brief description of the problem from a knowledge discovery perspective is presented.

Detecting objects and events in geoscience data is important for a number of reasons. For example, detecting spatial and temporal patterns in climate data can help in tracking the formation and movement of objects such as cyclones, weather fronts, and atmosphere rivers, which are responsible for the transfer of precipitation, energy, and nutrients in the atmosphere and ocean. Many novel pattern mining approaches have been developed to analyze the spatial and temporal properties of objects and events, e.g., spatial coherence and temporal persistence that can work with amorphous boundaries. One such approach has been successfully used for finding spatio-temporal patterns in sea surface height data, resulting in the creation of a global catalogue of Mesoscale Ocean eddies. The use of topic models has been explored for finding extreme events from climate time series data. Given the growing success of deep learning methods in mainstream machine learning applications, it is promising to develop and apply deep learning methods for a number of problems encountered in geosciences. Recently, deep learning methods, including convolutional neural networks (CNNs) and recurrent neural networks (RNNs) have been used to detect geoscience objects and events, such as detecting extreme weather events from a climate model.

Knowledge discovery methods can contribute to estimating physical variables that are difficult to monitor directly, e.g., methane concentrations in air or groundwater seepage in soil, using information about other observed or simulated variables. To address the combined effects of heterogeneity and small sample size, multi-task learning frameworks have been explored, where the learning of a model at every homogeneous partition of the data is considered as a separate task, and the models are shared across similar tasks. The sharing of learning is able to help in regularizing the models across all tasks and avoid the problem of over fitting. Focusing on the heterogeneity of climate data, online learning algorithms have been developed to combine the ensemble outputs of expert predictors and conduct robust estimates of climate variables such as temperature. To address the paucity of labeled data, novel learning frameworks such as semi-supervised learning, active learning, have huge potential to improving the state-of-the-art in estimation problems encountered in geoscience applications.

Forecasting long-term trends of geoscience variables such as temperature and greenhouse gas concentrations ahead of time can help in modeling future scenarios and devising early resource planning and adaptation policies. Some of the existing approaches in knowledge discovery for time-series forecasting include exponential smoothing techniques, the auto regressive integrated moving average model and probabilistic models, such as hidden Markov models and Kalman filters. In addition, RNN-based frameworks such as long-short-term-memory (LSTM) have been used for long-term forecasting geoscience variables.

An important problem in geoscience application is to understand the relationships in geoscience data, such as periodic changes in the sea surface temperature over the eastern Pacific Ocean and their impact on several terrestrial events such as floods, droughts and forest fires. One of the first knowledge discovery methods in discovering relationships from climate data is a seminal work, where graph-based representations of global climate data were constructed. In the work, each node represents a location on the Earth and an edge represents the similarity between the eliminated time series observed at a pair of locations. The high-order relationships could been discovered from the climate graphs. Another kind of method for mining relationships in climate science is based on representing climate graphs as complex networks, including approaches for examining the structure of the climate system, studying hurricane activity. Recently, some works have developed novel approaches to directly discover the relationships as well as integrating objects in geoscience data. For example, one work has been implemented to discover previously unknown climate phenomena. For causality analysis, the most common tool in the geosciences is bivariate Grange analysis, followed by multi-variate Granger analysis using vector auto regression (VAR) models.

8.2.2 Knowledge discovery methods

As a new strategic resource for human beings, big data has become a strategic highland in the era of knowledge economy. It is a typical representative of the data-intensive scientific paradigm following experience, theory and computational models, since this new paradigm mainly depends on data correlation to discover knowledge, rather than traditional causality. It is bringing about changes in scientific methodology, and will become a new engine for scientific discovery.

Knowledge discovery of remote sensing big data lies at the intersection of earth science, computer science, and statistics, and is a very important part of artificial intelligence and data science. Its aims at dealing with the problem of finding a predictive function or valuable data structure entirely based on data and will not be bothered by the various data types and, is suitable for comprehensively analyzing the Earth's big data.

The core knowledge discovery methods include supervised learning methods, unsupervised learning methods, and their combinations and variants. The most widely used supervised learning methods use the training data taking the form of a collection of (x, y) pairs, and aims to produce a prediction y' in response to a new input x' by a learned mapping f(x), which produces an output y for each input x (or a probability distribution over y given x). There are different supervised learning methods based on different mapping functions, such as decision forests, logistic regression, support vector machines, neural networks, kernel machines, and Bayesian classifiers. In recent years, deep networks have received extensive attention in supervised learning. Deep networks are composed of multiple processing layers to learn representations of data with multiple levels of abstraction, and discover intricate structures of the big earth data by learning its internal parameters to compute the representation in each layer. Deep networks have brought about breakthroughs in processing satellite image data, forecasting long-term trends of geoscience.

Unlike supervised learning methods, unsupervised learning involves the analysis of unlabeled data under assumptions about structural properties of the data. For example, the dimension reduction methods make some specific assumptions that the earth data lie on a low-dimensional manifold and aim to identify that manifold explicitly from data, such as principal components analysis, manifold learning, and auto encoders. Clustering is another very typical unsupervised learning algorithm, which aims to find a partition of the observed data, and mine the inherent aggregation and regularity of data. In recent years, much current research involves blends across supervised learning methods and unsupervised learning. Semi supervised learning is a very typical one, which makes use of unlabeled data to augment labeled data in a supervised learning context considering the difficulty of obtaining some geoscience supervision data. Overall, knowledge discovery of the big earth data needs to leverage the development of artificial intelligence, machine learning, statistical theory, and data science.

8.2.3 Related work

With the growing numbers of archived RS images for Earth observation, an increasing number of scientists are interested in the spatiotemporal analysis of RS data. Many researchers proposed combining online analytical processing(OLAP)[245] technology with the GIS[246] to build a data cube. They built the multidimensional database paradigm to manage several dimension tables, periodically extracting the dimension information from the data in GIS, and achieved the ability to explore spatiotemporal data using the OLAP space–time dimension aggregation operation. Sonia Rivest et al. deployed a spatial data warehouse based on GIS and spatial database to acquire, visualize, and analyze the multidimensional RS data[247]. Matthew Scotch et al. developed the SOVAT tool[248], using OLAP and GIS to perform the public health theme analysis with the data composed of spatiotemporal dimensions. These tools can facilitate researchers extracting data with spatiotemporal dimensions; however, their multidimensional data model is unsuitable for complicated scientific computing. Further, they did not adopt an appropriate architecture for large data processing[249][250]. Therefore, their ability to handle large-scale data is limited.

Owing to natural raster data structure of Earth observation images, the time-series imagery set can be easily transformed to multidimensional array. For example, a 3D array can represent the data with spatiotemporal dimensions. This data type is suitable for parallel processes, because a large array can be easily partitioned into several chunks for distributed storing and processing. In addition, the multidimensional array model enables a spatiotemporal auto-correlated data analysis; therefore, researchers need not be concerned about the organization of discrete image files. Thus, much research is focused on developing new analysis tools to process the large RS data based on the multidimensional array model; e.g., Gamara et al.[251] tested the performance of spatiotemporal analysis algorithms on array database architectures - SCIDB[252], which described the efficiency of spatiotemporal analysis based on the multidimensional array model, Assis et al.[237] built a parallel RS data analysis system based on the MapReduce framework of Hadoop[253], describing the 3D array with key/value pairs. Although these tools have significantly improved the computation performance of RS data analysis, they also contain some deficiencies. First, many of them focused only on analyzing the RS raster image data located by geographic coordinates, and did not provide the support of spatial feature, thereby limiting their ability to use these geographic objects in the analysis application. Next, some of these tools require analysers to fit their algorithms into specialised environments, such as Hadoop MapReduce framework[254]. This will be user unfriendly to researchers who only desire to focus on their analysis application.

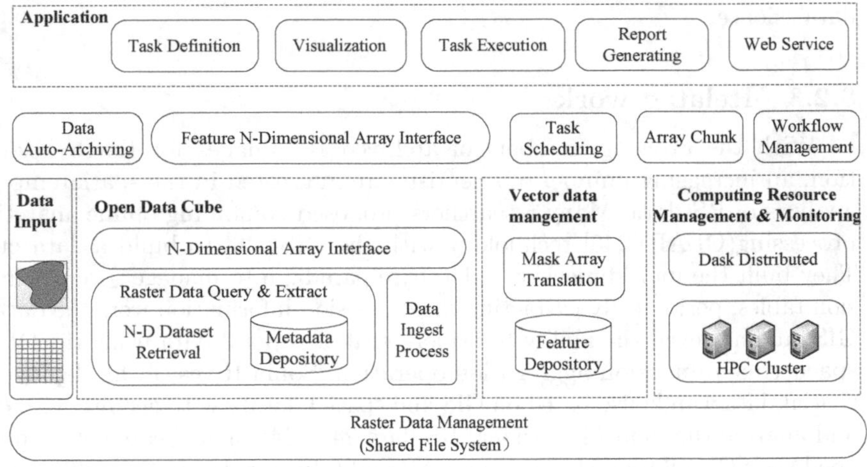

FIGURE 8.1: The architecture of the SRSDC.

8.3 Architecture Overview

8.3.1 Target data and environment

The SRSDC system is designed for providing the services of large RS data time-series analysis with spatial features, and it aims to manage and process the large-scale spatial feature data and satellite images seamlessly. Based on the open data cube(ODC)[255], which is a popular data cube system used for spatial raster data management and analysis, we archived large amounts of satellite data within China. These data came from different satellites including Landsat, MODIS, GaoFen(GF) and HuanJing(HJ). In addition, the SRSDC also contains many features data within China, such as lakes, forests and cities. These spatial vector data were downloaded from the official web site of OpenStreetMap[256]. Before obtaining these satellite images in the SRSDC, the geometric correction and radiometric correction for these images must be ensured. This can ensure the comparability between the images in different time, space and measurements; subsequently, the global subdivision grid can be used to partition the data into many tiles(grid files). These tiles were stored as the NetCDF format[257], which supports many analysis libraries and scientific toolkits.

8.3.2 FRSDC architecture overview

The SRSDC system adopted the relational database and file system to manage the spatial data. It is designed to be scalable and efficient and provide feature support for time-series analysis. Compared with the ODC system[255], which only supports spatial raster data management and analysis as a data cube, the SRSDC supports the extraction of target satellite data as a multi-dimensional array with the geographic object. It could perform the spatial query operation with geographic objects, instead of locating data with only geographic coordinates. Therefore, the dataset built for analysis has geographic meaning. Thus, if researchers desire to obtain the target dataset, they only need to query data with the geographic meaning of the analysis themes, without knowing specific geographic coordinates. As shown in Figure 8.1, the system is primarily composed of the data management and distributed execution engine(DEE). Data management consists of two parts, raster data management and vector data management. For the raster data, the SRSDC will archive it into a shared network file system and extract its metadata information automatically; these metadata will be stored into the metadata depository managed by ODC. For the vector data, the SRSDC stores them as geographic objects in the spatial database. After the data management, an N-Dimensional array interface is responsible for transforming the raster data and vector data to an N-Dimensional array that has the spatial feature information. Xarray[258] is used for array handing and computing. DEE is responsible for providing the computing environment and resources on high performance computing(HPC) clusters. The SRSDC use dask[244], which is a parallel computation library with blocked algorithms, for the task scheduling, distributed computing, and resource monitoring. It could help researchers to execute the analysis tasks in parallel.

8.4 Design and Implementation

8.4.1 Feature data cube

8.4.1.1 Spatial feature object in FRSDC

Spatial feature is a geographic object that has special geographic meaning. It is often important for RS application, because researchers occasionally need to process the RS image dataset with geographic objects, such as the classification of an RS image with spatial features[259][260][261]. However, many RS data cube systems only provide the multidimensional dataset without features. Hence, researchers are required to perform additional work to prepare the data for analysis. For example, the ODC system[255] and the data cube based on SCIDB[251] could only query and locate the study area by simple geographic coordinates, so researchers must transform their interested spatial features to coordinate ranges one by one if they want to prepare the analysis ready data. To solve this problem, the SRSDC combined the basic N-Dimensional array with geographic objects to provide the feature N-Dimensional array for researchers, and researchers could easily organise the analysis ready N-Dimensional dataset by their interested features. Within the SRSDC, now we primarily archive the forest and lake features of China, and store them as geographic objects in a PostGIS database. The unified modeling language(UML) class diagram in the SRSDC is shown in Figure 8.2, and the description of these classes is as follows:

1. The feature class is provided for users to define their spatial feature of interest with a geographic object. It contains the feature type and the geographic object.

2. The feature type class represents the type of geographic object, such as lakes, forests, cities and so on. It contains the description of the feature type and analysis algorithm names that are suitable for the feature type.

3. The geographic object class describes the concrete vector data with geographic meaning, such as Poyang lake (a lake in China). It contains the vector data type to illustrate its geometry type.

4. The raster data type class is used to describe the type of satellite data. It contains the satellite platform, product type, and bands information.

5. The feature operation class is used to extract the feature datacube dataset from the SRSDC. It contains the feature object, raster data type, and time horizon to build the target feature N-Dimensional array. It also provides some operation functions for users.

8.4.1.2 Data management

As mentioned above, data management consists of raster data management and vector data management. Because of satellite data's large volume and variety, the SRSDC uses the file system to store the raster data and manage the metadata by a relation database that contains NoSQL fields. In the

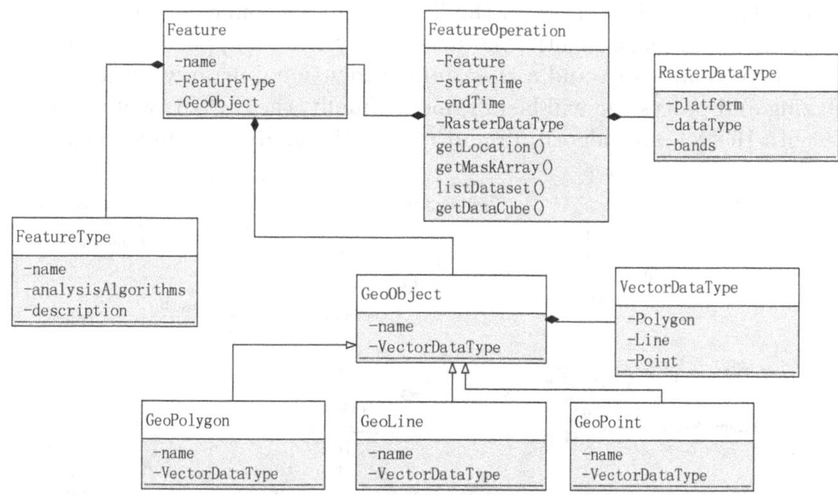

FIGURE 8.2: The UML class diagram in the SRSDC.

metadata depository, the SRSDC uses NoSQL fields to describe the metadata information instead of a full relation model. This is because the number of satellite sensors is increasing rapidly, and if the full relation model is used, the database schema must be expanded frequently to meet the new data sources. In contrast, NoSQL fields are more flexible in describing the metadata of satellite data that originate from different data sources. The NoSQL fields contain the time, coordinate, band, data URL, data type, and projection. Among these fields, some are used for data query, such as the time, band, coordinates, and data type. Some other fields are used for loading the data and building the multidimensional array in memory, such as data URL and projection. In addition, comparing the vector data volume(GB level) we download from the OpenStreetMap[256] with the raster data volume(TB level) we archived from several satellite data centers[262], we found that most spatial features that are represented by the vector data are not as large as the raster images; therefore we established a feature depository instead of file system to store and manage them as geographic objects. These objects may contain different geographic meanings, and we defined them as feature types.

The runtime implementation of feature data cube building and processing is shown in Figure 8.3. First, the SRSDC receives the user's data request from the web portal and obtains the target geographic object by querying the feature depository. Next, it conducts a feature translation to transform the geographic object into a mask array and obtains the minimum bounding rectangle(MBR) of the feature. Subsequently, with the vertex coordinates of MBR and time horizon, the SRSDC searches for the required raster data's metadata to locate physical URLs of the raster data. Next, ODC's N-Dimensional array interface will load the raster data set from the file system and build a multidimensional array in memory. Subsequently, the mask array will be applied to masking the multidimensional array, and a new multidimensional array with features for analyzing and processing will be obtained. Finally, the SRSDC will process the data with the relevant algorithm and return the analysis results to the user.

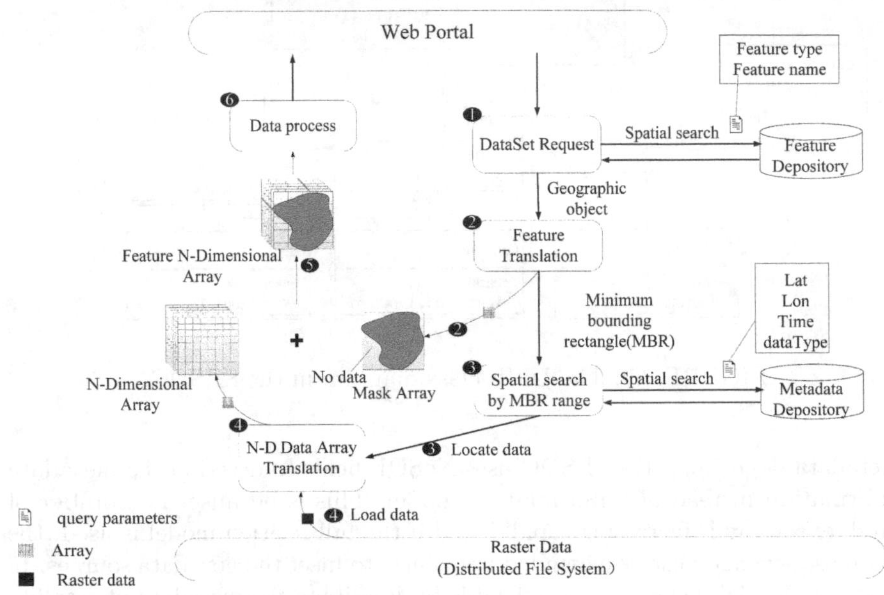

FIGURE 8.3: Runtime implementation of feature data cube building.

8.4.2 Distributed executed engine

As an increasing number of RS applications need to process or analyze the massive volume of RS data collections, the stand-alone mode processing can not satisfy the computation requirement. To process the large-scale RS data efficiently, we built a distributed executed engine using the dask – a distributed computing framework focusing on scientific data analysis. Compared with the popular distributed computing tools such as Apache Spark, dask supports the multidimensional data model natively and has a similar API with pandas and numpy. Therefore, it is more suitable for computing an N-Dimensional array. Similar to Spark, dask is also a master-slave system framework that consists of one schedule node and several work nodes. The schedule node is responsible for scheduling the tasks, while the work nodes are responsible for executing tasks. If all the tasks have being performed, these workers' computation results would be reduced to the scheduler and the final result would be obtained.

In the SRSDC, we could index the satellite image scenes by adding their metadata information to the database, and then obtain the data cube dataset

(N-Dimensional arrays) from the memory for computing. However, to compute the large global dataset, we should slice the large array into the fixed-size sub-arrays called chunks for computing in the distributed environment. The SRSDC partitions these native images into seamless and massive tiles based on a latitude/longitude grid. The tile size is determined by the resolution of satellite images. For example, in the SRSDC, the Landsat data(each pixel 0.00025°) was partitioned into tiles of size 1°x 1°, and the tiles(4000x4000 pixels array) can be easily organized as a data chunk, which is suitable for the memory in the worker node. By configuring the grid number and time horizon, the chunk could be built. Further, with these data chunks, the SRSDC can transform the big dataset(N-Dimensional arrays) to several sub-arrays loaded by different worker nodes. After all the data chunks have been organized, the scheduler will assign the chunks to the workers and map the functions for computing.

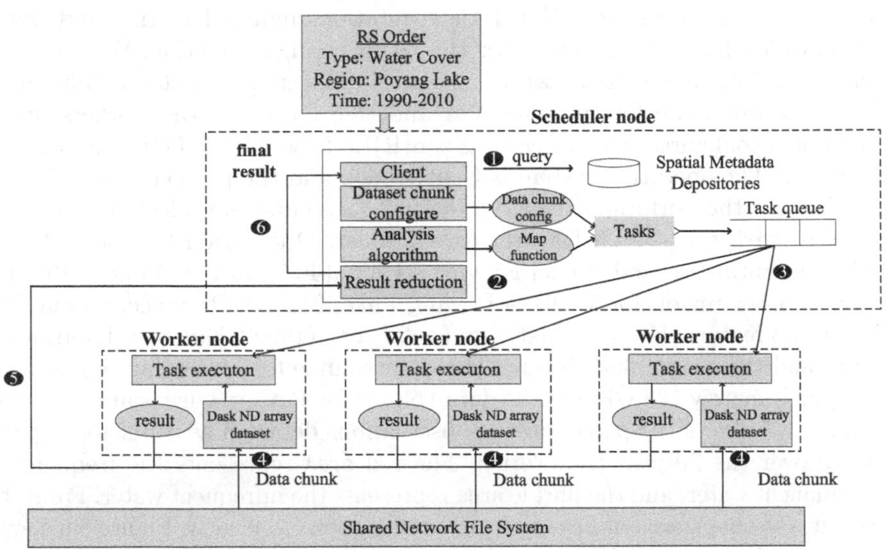

FIGURE 8.4: Large-scale RS analysis processing with distributed executed engine.

As shown in Figure 8.4, the processing of large-scale time series analysis by the distributed executed engine is as follows:

1. Organize the data cube dataset by multidimensional spatial query.

2. Configure the appropriate parameters(grid number or time horizon) to organized the data chunks for workers, manage the chunks' ids with a queue.

3. Select the analysis algorithm and data chunks to compose the tasks, and assign these tasks to the worker node.

4. Check the executing state of each task in the workers; if failure occurs, recalculate the result.

5. Reduce all the results to the scheduler and return the analysis result to the client.

8.5 Experiments

To verify the ability of multidimensional data management and large-scale data analysis in the SRSDC, we conducted the following time-series analysis experiments focusing on spatial feature regions and compared the performance of GEE and stand-alone mode processes on the target dataset.

In this experiment, two RS application algorithms for time-series change detection have been used: NDVI for vegetation change detection and water observation from space(WOfs) for the water change detection. We built the distributed executed engine with four nodes connected by a 20 GB Infiniband network; one node for the scheduler and tree nodes for the workers. Each node was configured with Inter(R) Xeon(R)E5-2460 CPU(2.0GHz) and 32GB memory. The operating system is CentOS 6.5, and the python version is 3.6.

To test the performance of the feature data cube, we selected two study regions with the special features as examples. One region for the NDVI is Mulan hunting ground, Hebei Province, China(40.7°-43.1°N, 115.8°-119.1°E), and another region for WOfs is Poyang Lake, Jiangxi Province, China. We built two feature data cube datasets for 20 years(1990-2009) with Landsat L2 data and the geographic objects. The data volume for Mulan hunting ground is approximately 138 GB and the data volume for Poyang Lake is about 96 GB. Figure 8.5 shows the percentage of observations detected as water for Poyang Lake over the 20-year time series. The red area represents the frequent or permanent water, and the purple area represents the infrequent water. From the result, the shape area of Poyang Lake can be observed clearly. Figure 8.6 shows the annual average NDVI production on the Mulan hunting ground; Figure 8.7 shows the NDVI time series result of the sampling site(41.5620°N, 117.4520°E) over 20 years. As shown, the values during 2007-2008 were abnormally below the average. This is because the average annual rainfall during this time is lower than that in normal years.

To test the processing performance of the DEE for different amounts of data, we tested time consumed by processing 6.3 GB, 12.8 GB, 49.6 GB, 109.8 GB, 138.6GB input data for NDVI production. These data have been partitioned into 4000x4000 pixels tiles mentioned above, with which we compared the performances of the stand-alone model and DEE models:

1. stand-alone model: organize the dataset as data chunks, and process these data chunks serially with a single server.

2. DEE model: organize the large dataset as data chunks, and assign

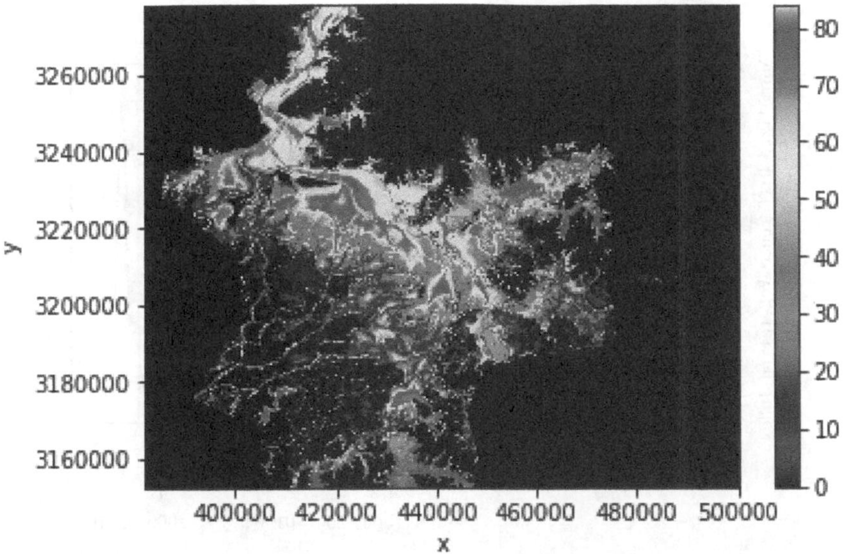

FIGURE 8.5: Water area of Poyang Lake over 20-year time series.

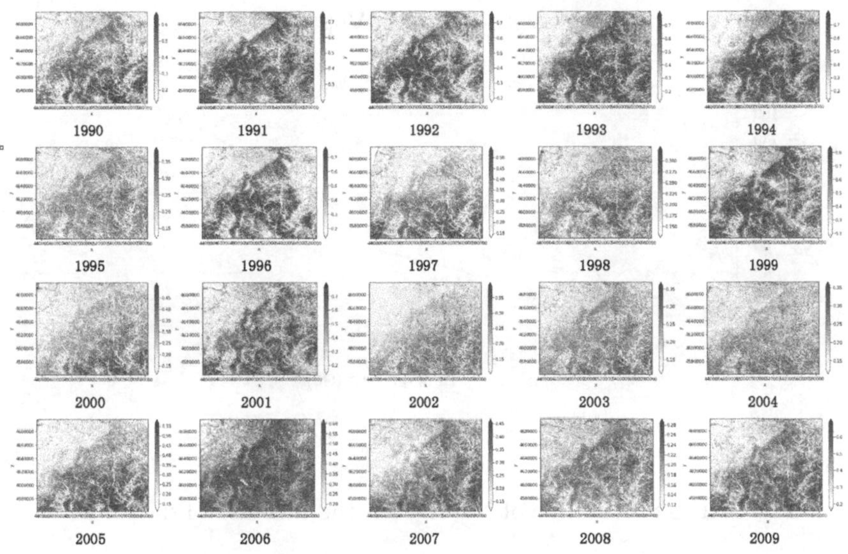

FIGURE 8.6: The annual average NDVI of Mulan hunting ground for 20 years.

different workers to read these data chunks to process them in parallel with the distributed executed engine, which consists of one schedule node and three work nodes.

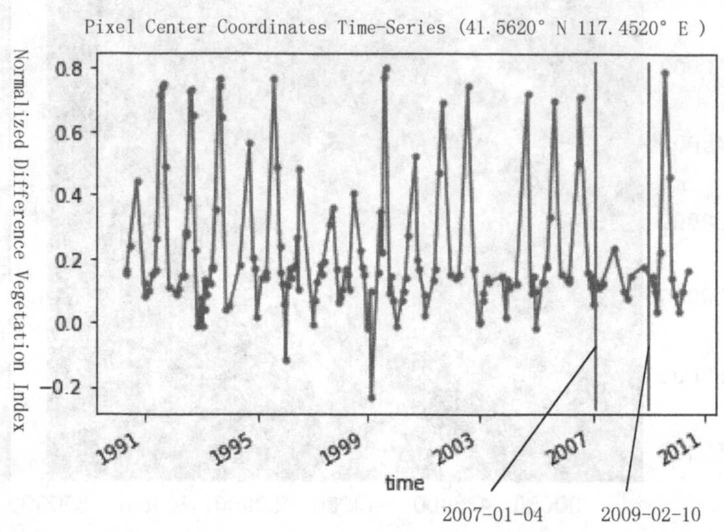

FIGURE 8.7: A NDVI time series of sampling site on Mulan hunting ground.

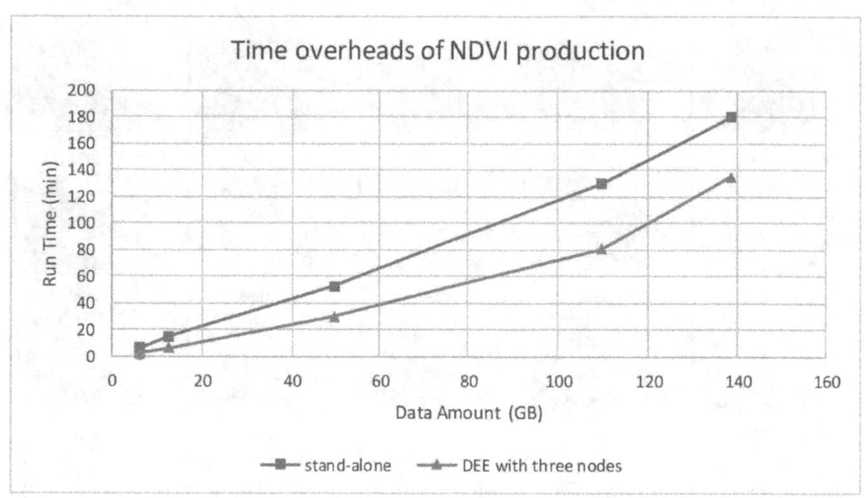

FIGURE 8.8: Runtime of NDVI with the increase of data volume.

As shown from the experimental results in Figure 8.8, the DEE mode is much faster than the stand-alone mode because it can use the shared memory of clusters nodes and process the large dataset in parallel. As the process data amount increases, we also observed that the time consumed will grow nonlinearly. This is due to the IO limit of the shared network file system and scheduling overhead. The speedup performance when generating the NDVI

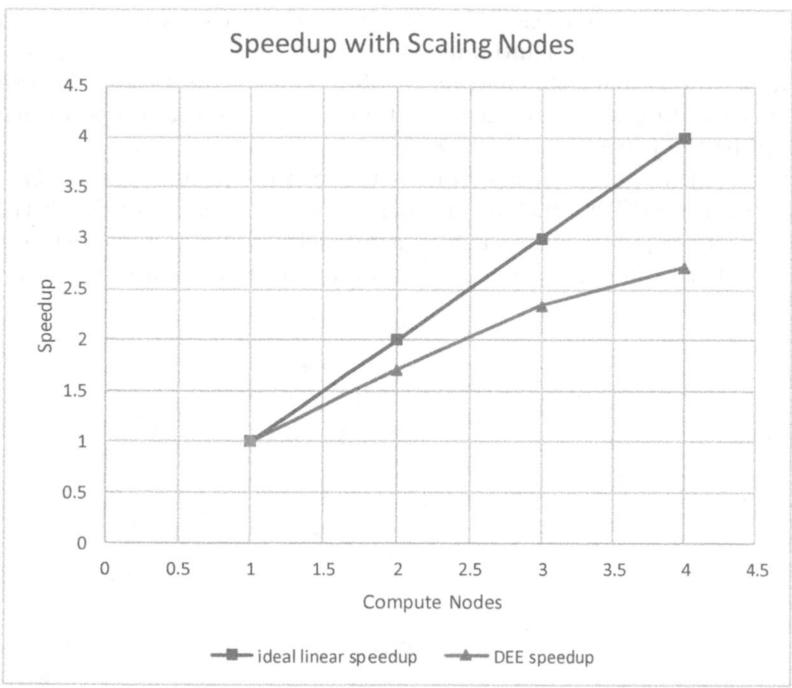

FIGURE 8.9: Speedup for the generation of NDVI products with increasing nodes.

production with increasing numbers of work nodes also proved this point, as shown in Figure 8.9. Therefore, we conclude that the SRSDC has a certain capacity to process the massive data, which is unsuitable for the memory.

8.6 Conclusions

We have designed and tested a feature supporting, scalable, and efficient data cube for time-series analysis application, and used the spatial feature data and remote sensing data for comparative study of the water cover and vegetation change. In this system, the feature data cube building and distributed executor engine are critical in supporting large spatiotemporal RS data analysis with spatial features. The feature translation ensures that the geographic object can be combined with satellite data to build a feature data cube for analysis. Constructing a distributed executed engine based on dask ensures the efficient analysis of large-scale RS data. This work could provide a convenient and efficient multidimensional data services for many remote sensing applications[263][264]. However, it also has some limitations; for example, the image data is stored in the shared file system, and its IO performance is limited by the network.

In the future, more work will be performed to optimize the system architecture of the SRSDC, such as improving the performance of the distributed executed engine, selecting other storage methods which could ensure the process data locality, adding more remote sensing application algorithms, etc.

Chapter 9

Automatic Construction of Cloud Computing Infrastructures in Remote Sensing

9.1 Introduction

With the advent of the information age, the high-spectral resolution and high-spatial resolution sensors are launched, and the remote sensing data obtained by satellite ground stations has exploded exponentially. The acquisition and update cycle of remote sensing data is greatly shortened, and the timeliness has become increasingly stronger [265, 266]. How to process effectively these high-scoring remote sensing data becomes a huge challenge to the traditional spatial information system in terms of computing, storage, and data transmission.

The construction of a traditional remote sensing information system is that remote sensing information service providers offer services. Remote sensing information users build their own software and hardware systems, deploy remote sensing information applications, and invoke resources provided by the service providers [267]. This traditional way of constructing a remote sensing information system causes many problems, such as high cost of system construction and difficulty of system expandability, etc. In order to solve these problems, cloud computing is introduced into the field of remote sensing, and some remote sensing oriented cloud computing platforms have been established. Compared with the traditional remote sensing information system, the remote sensing cloud platform makes full use of resources, and provides remote sensing data services for large-scale applications, providing users with on-demand and personalized products. Otherwise, the technology research on cloud computing-based resource virtualization, develops the remote sensing oriented cloud computing environment that meets the needs of multiple users [268, 269].

In addition, the remote sensing cloud platform system architecture is complex, the system deployment is difficult and the cost is high, and some professional remote sensing software installation processes are cumbersome and complicated. In response to these problems, this study realizes one-click efficient deployment of the remote sensing cloud platform and remote sensing softwares by researching the automatic deployment technology SaltStack.

It can be said that the development of remote sensing oriented cloud computing is still immature. In order to effectively play the role of cloud computing in the field of remote sensing, respond to the challenges of remote sensing big data and fully play the huge advantages of cloud computing, relevant research on infrastructure automatic construction of cloud computing services in remote sensing becomes very necessary.

9.2 Definition of the Remote Sensing Oriented Cloud Computing Infrastructure

In general, cloud computing includes the following levels of service: Software as a Service (SaaS), Platform as a Service (PaaS), and Infrastructure as a Service (IaaS) [270, 271]. The so-called hierarchy here is the "level" in the sense of the platform architecture. IaaS, SaaS, and PaaS are implemented at the infrastructure layer, the application software layer respectively, and the software running platform layer.

The infrastructure in cloud computing mainly utilizes virtualization technology to virtualize the physical infrastructure into virtual machine computing resources, and provide services to users through the Internet in the form of virtual resource pools [272, 273].

9.2.1 Generally used cloud computing infrastructure

In the common cloud computing infrastructure architecture, the cloud computing infrastructure corresponds to an abstraction layer of a virtualized resource pool; the virtual resource pool is hosted by the service providers, and delivered to the users through the Internet [27].

The generally used cloud computing infrastructure mainly includes three aspects: cloud infrastructure components, cloud infrastructure architecture and cloud infrastructure services.

Cloud infrastructure components mostly refer to back-end hardware devices that support the operation of the system platform, including computing servers, storage devices, and network devices.

There are three main types of cloud computing infrastructure architecture: private, public, and hybrid. The private cloud model represents that users purchase and build cloud infrastructure components themselves. The public cloud model means that users rent cloud infrastructure and cloud infrastructure services provided by third-party cloud providers on demand via the Internet. The hybrid cloud model merges the private cloud model and the public cloud model.

Cloud infrastructure services (IaaS) is a cloud model that enables users to rent IT infrastructure components (including computing, storage, and networking) over the Internet [274]. This public cloud service model is often referred to as IaaS. IaaS eliminates up-front capital costs associated with the on-premises infrastructure and follows a usage-based consumption model. In this pay-per-use model, users pay only for basic consumption of services, usually in hours, days, and months [275]. Cloud computing providers typically price IaaS in a metered manner, with rates corresponding to a given level of performance. For virtual servers, this means that the various server sizes correspond to different prices, usually measured in terms of standard virtual CPU sizes and memory increments [276, 277]. Currently major IaaS vendors include AWS, Google, Microsoft, IBM, Alibaba.

9.2.2 Remote sensing theme oriented cloud computing infrastructure

In the field of remote sensing, the core goal of cloud computing infrastructure is to provide services in a bundled manner between remote sensing elements and computer resource elements. Cloud computing infrastructure is inherently highly scalable and does a better job of scaling strategies than traditional physical hardware. Remote sensing information processing and remote sensing data storage have relatively large requirements on computer performance. Cloud computing infrastructure can integrate physical machine resources through virtualization technology, thus providing a remote sensing computing environment that can meet computing and storage needs of cloud users. Affected by remote sensing oriented big-data, more flexible cloud computing infrastructure services are becoming more necessary in the field of remote sensing.

The application of remote sensing oriented cloud computing infrastructure can be divided into three layers:

- The bottom layer is the infrastructure resource layer, which uses virtualization technology to eliminate the differences of heterogeneous physical resources, and integrates computing, storage and network resources into a unified virtual resource pool;
- The middle layer is the infrastructure management layer, which mainly designs services such as load balancing, virtual resource monitoring, information security, resource dynamic optimization, and system data backup so as to ensure the high availability, high stability and high quality of cloud computing infrastructure services;
- The top layer is the service interface layer, which provides remote sensing cloud services through unified interfaces, including computing, storage, network and other services.

Due to the huge advantages of developing cloud computing, remote sensing oriented cloud computing infrastructure technologies have developed particularly rapidly in recent years. Many excellent remote sensing cloud platform products have been developed. OpenRS-Cloud [278], an open remote sensing data processing and service platform, has been developed by Wuhan University. It can store and process time-space remote sensing data with structural sparsity [279], and has multiple land cover classification remote sensing technologies [280, 259]. SuperMap GIS6R [281], one customized GIS cloud computing platform, is launched by the US Environmental Systems Research Institute. The GIS Cloud platform is based on cloud computing technologies, which provides support for the space information system by taking advantage of the main features of cloud computing. The architecture of the cloud GIS platform mainly makes full use of various features of cloud computing to improve the traditional GIS system architecture, change the traditional GIS application methods and construction methods, and make the computing power and data processing capacity more efficient. The key technologies of the GIS cloud platform architecture mainly include: cloud storage technologies of massive spatial data, virtualization technologies, GIS automated deployment and load balancing technologies.

9.3 Design and Implementation of Remote Sensing Oriented Cloud Computing Infrastructure

The requirements of the remote sensing cloud can be specifically divided into two aspects. On the one hand are, the requirements of data, such as the acquisition, storage, retrieval and download of multi-source heterogeneous remote sensing data. On the other hand, there are requirements for remote sensing data processing, including providing cloud software for processing remote sensing data and cloud system platform for running remote sensing software.

The remote sensing oriented cloud computing infrastructure provides a virtual computing environment, which is mainly applied to the remote sensing computing platform, which solves the key problems of high deployment cost, difficulty in deployment and transplantation, and serious waste of resources. It is necessary to provide a low-cost, convenient and highly scalable remote sensing oriented computing platform for users who need to handle remote sensing data. At the same time, the remote sensing computing platform corresponding to the cloud infrastructure also provides users with resource monitoring services; that is, the platform provides users with network status monitoring, node resource monitoring (including CPU usage, memory usage, disk usage) and calculation task monitoring, etc.

9.3.1 System architecture design

The architecture of the remote sensing oriented cloud computing system platform is mainly divided into four aspects: resource layer, management layer, computing layer and service layer.

The remote sensing oriented cloud computing infrastructure is applied to the computing platform in the remote sensing cloud system, and its design is mainly embodied in the resource layer and management layer of the overall architecture of the remote sensing oriented cloud system platform. The computing platform implemented by the remote sensing oriented cloud computing infrastructure can be subdivided into three layers: infrastructure resource layer, infrastructure management layer and unified service interface layer. The framework of remote sensing oriented cloud infrastructure is shown in Figure 9.1.

- The infrastructure resource layer includes not only basic physical resources of computing, storage and network, but also remote sensing resources such as satellite data, geographic data, remote sensing software, remote sensing data processing algorithm library, etc.

- The infrastructure management layer, including cloud computing platform framework OpenStack and automated deployment tool SaltStack, is responsible for the management of cloud computing infrastructure.

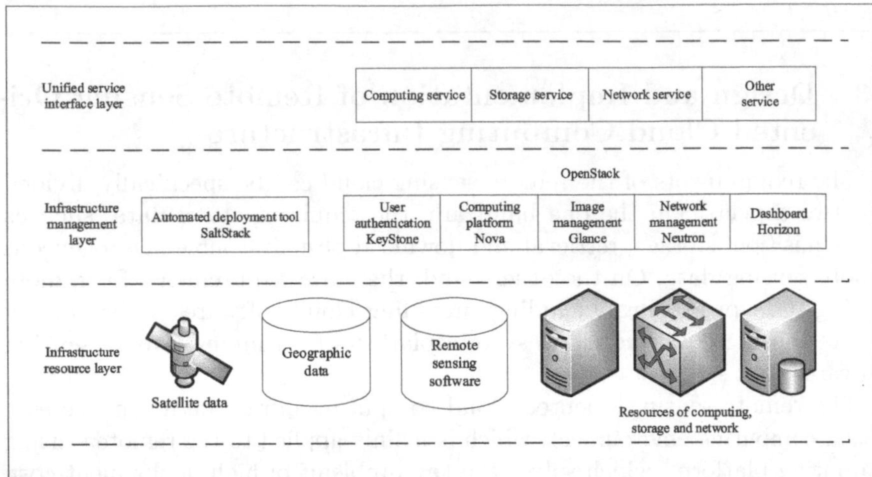

FIGURE 9.1: Framework of remote sensing oriented cloud infrastructure.

- The unified service interface layer includes computing service, storage service, and network service, etc.

9.3.2 System workflow design

The remote sensing oriented cloud system platform mainly includes the data platform and the computing platform. This section mainly introduces the workflow design of the computing platform corresponding to the cloud computing infrastructure.

The remote sensing cloud computing platform corresponding to the cloud infrastructure is based on two open source softwares, OpenStack and SaltStack. Its workflow is summarized as the following four main steps:

(1) Tenants submit platform orders (that is, remote sensing virtual environment configuration, including cloud host system selection, host basic configuration and remote sensing softwares installation);

(2) Analyzing and processing the orders in the background;

(3) The remote sensing cloud system background deploys the platform environment according to the parameter information parsed by the orders;

(4) New functions can be developed for the ordered remote sensing virtual platform.

The workflow of a remote sensing oriented cloud computing platform is shown in the following Figure 9.2.

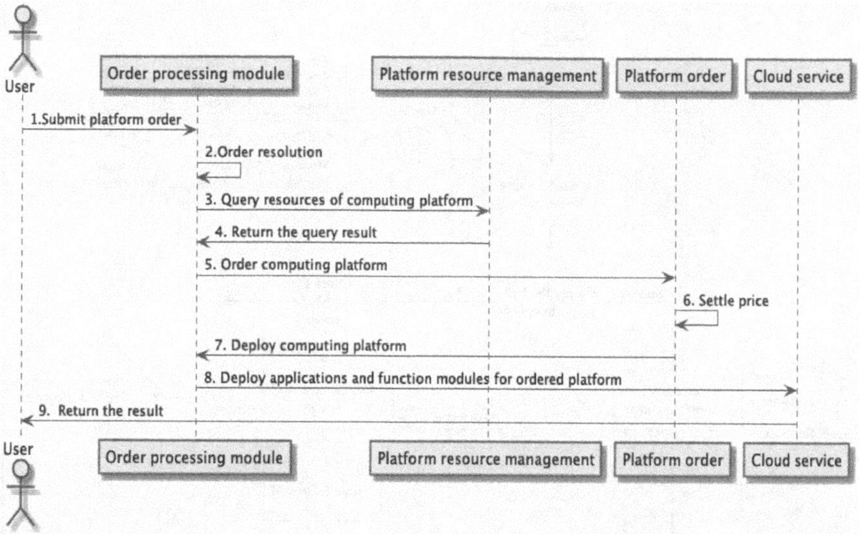

FIGURE 9.2: Workflow of remote sensing oriented cloud computing platform.

In the design of a remote sensing oriented cloud computing platform, the platform order processing workflow is the deployment process of the remote sensing cloud computing virtual environment (Virtual environment, VE) based on OpenStack architecture:

(1) Generating order. The user selects the computing environment VE that needs to be ordered, and the system generates an XML description file according to the user's order selection.

(2) The legality test of resource requirements. The system receives the user's resource requirement XML file, so as to verify the legality of each resource requirement.

(3) Query the VE image library. According to the XML file of the requirement, the VE image library is indexed. If there is an image of the virtual computing environment that meets the conditions, the image ID is returned, and the next step of production is performed. If it does not exist, the system image is not present and the operation is exited.

(4) Query the software library. The software package resource is retrieved from the software repository and the software package ID is returned.

(5) Specific deployment. According to the resource information returned in the previous two steps, the image is deployed on the OpenStack platform to produce a customized virtual computing environment.

The automatic order processing workflow in the remote sensing oriented cloud computing platform is shown in the following Figure 9.3.

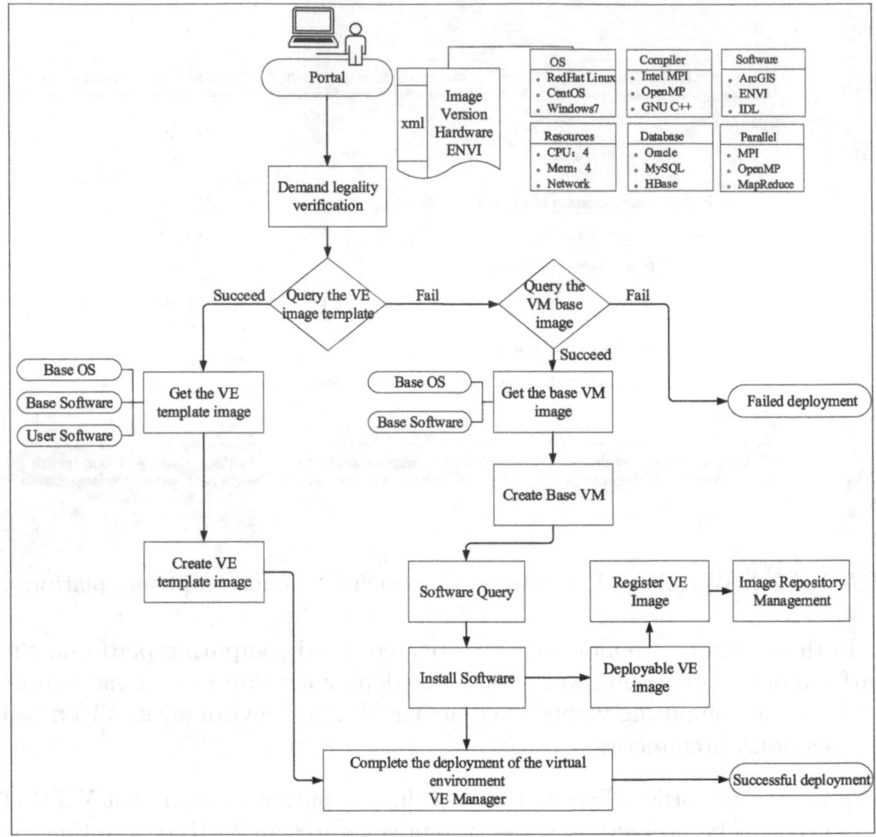

FIGURE 9.3: Workflow of order processing in remote sensing oriented cloud.

9.3.3 System module design

The remote sensing oriented cloud system computing platform mainly provides function modules for tenants as shown in Figure 9.4.

- Remote sensing data: realizing standard remote sensing data and remote sensing value-added products of searching, browsing, ordering, downloading and transferring services for remote sensing cloud tenants.
- Remote sensing production: realizing online production of remote sensing value-added products for remote sensing cloud tenants.
- Remote sensing computing software: providing remote sensing cloud tenants with the required remote sensing computing software.
- Remote sensing computing platform module: providing remote sensing virtual computing environment for remote sensing cloud tenants, including computing power and storage capacity. The remote sensing cloud

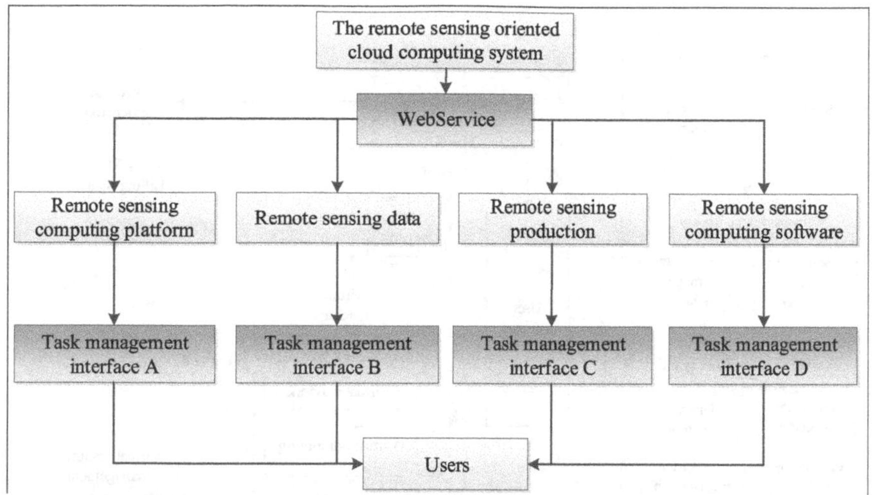

FIGURE 9.4: Functions of remote sensing oriented cloud computing platform.

computing platform service is part of the cloud infrastructure layer (IaaS) construction in the remote sensing cloud system.

The remote sensing oriented cloud infrastructure is mainly applied to the remote sensing computing platform module. The module includes three sub-function modules: computing resource service, storage resource service and cloud resource monitoring management.

Both computing and storage resources are cloud infrastructure provided by remote sensing cloud systems. The functions of this part include system image library management, software image database management, remote sensing computing platform customization, virtual resource management and so on.

The detailed functions of the computing platform service module (shown in Figure 9.5) in the remote sensing cloud system are as follows:

- System image library management: This part is mainly responsible for managing thc system image that the cloud host needs to use. The remote sensing cloud tenant can query the existing system image through database query, or apply for uploading the system image that you need to personalize production.

- Software image library management; this part not only manages the remote sensing software contained in system software warehouse, but also provides the function of the remotc sensing cloud tenant to upload remote sensing software. Once successfully registered, users can upload their own customized remote sensing software to the cloud host.

- Customization of remote sensing computing platform; this part is mainly based on the personalized selection of remote sensing cloud tenants,

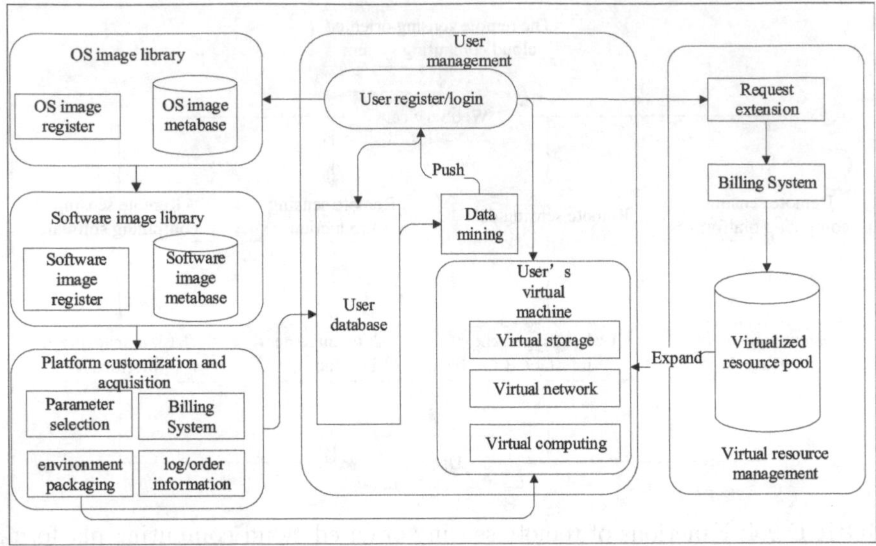

FIGURE 9.5: The detailed functions of the computing platform service module.

so as to deploy cloud hosts and build the remote sensing environment (including remote sensing algorithm library compilation and installation, remote sensing software installation and configuration, and construction of virtual cluster).

- Virtual resource management; this part is mainly for the remote sensing cloud tenants to manage their own customized remote sensing virtual computing environment, including expansion of scale, cloud host renewal and other services.

9.4 Key Technologies of Remote Sensing Oriented Cloud Infrastructure Automatic Construction

9.4.1 Automatic deployment based on OpenStack and SaltStack

In the field of remote sensing, the automatic deployment of the remote sensing cloud environment includes two main problems: one is the installation and configuration of remote sensing software, and the other is the virtualization of the computer hardware environment. The main problems faced by the software for automatic installation are the complex remote sensing data processing algorithms, the cumbersome installation steps of remote sensing data processing software, the high price of large-scale remote sensing commercial software, and the difficulty in sharing and multiplexing between different projects. Aiming at the hardware environment, it is mainly how to realize the virtualization of computing resources, storage resources and network resources [282, 283, 284].

Combined with open source cloud computing architecture OpenStack and heterogeneous platform infrastructure management tool SaltStack, the cloud infrastructure platform in the remote sensing cloud environment is automatically built, and the virtual environment with remote sensing service capability is automatically constructed [285]. WebService offers users convenient and one-click software installation services.

The overall architecture of OpenStack is shown in Figure 9.6.

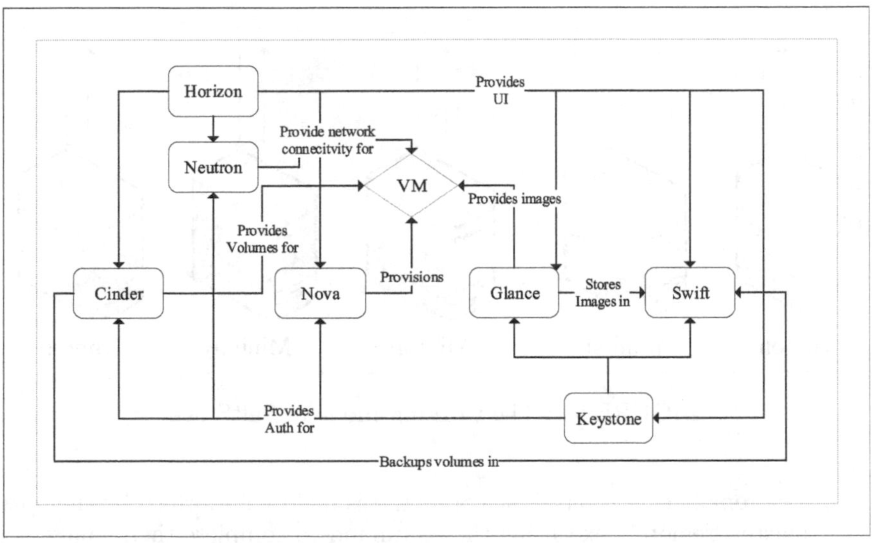

FIGURE 9.6: The overall architecture of OpenStack.

SaltStack is a heterogeneous platform infrastructure management tool, also known as an automatic operation tool. SaltStack supports most operating systems including Linux, Windows, and Solaris, allowing administrators to perform unified operations on different operating systems and achieve second-level communication between servers. SaltStack is one C/S architecture that uses the lightweight communicator ZMQ [286]. It is a batch management tool written in the Python programming language. It is completely open-source, complied with the Apache2 protocol, and similar to Puppet and Chef. It has a remote execution command engine and a configuration management system, called the Salt State System. In the configuration of SaltStack, the server is called master, the client is called minion; the server master and the client minion communicate through the ZeroMQ message queue, which can implement configuration management for the client minion. The working mode of SaltStack is shown in Figure 9.7.

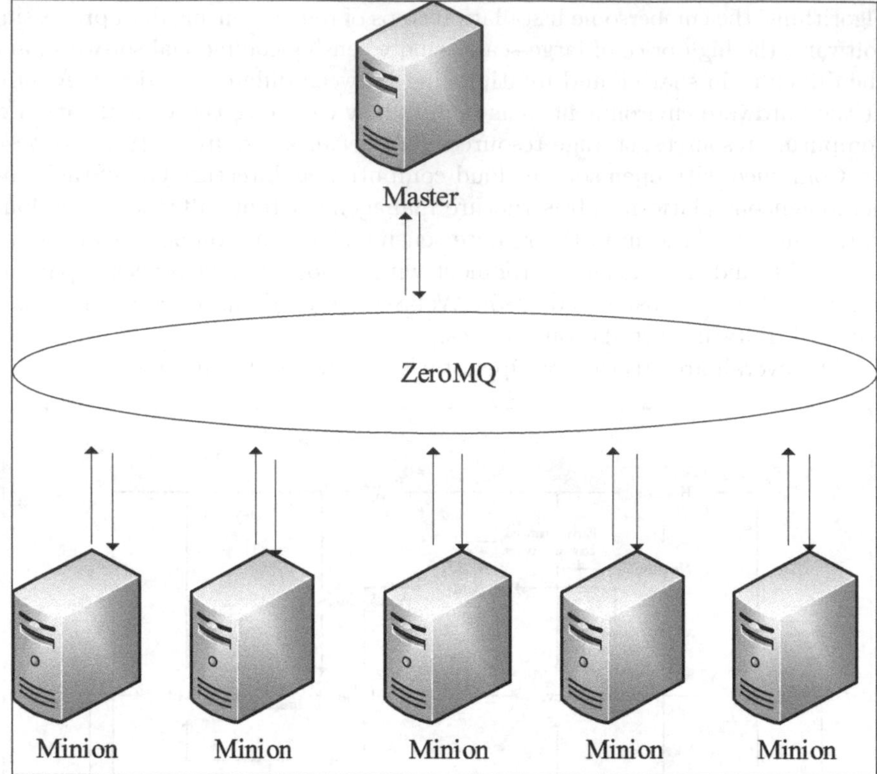

FIGURE 9.7: The working mode of SaltStack.

(1) The remote execution of the command is mainly performed by the server master, remotely executing the command to complete the management client minion work, so that the administrator performs the corresponding command operation without logging in to the managed node, and obtains the execution result. Among them, the remote execution function mainly uses the SaltStack Module (the management module integrated for a series of execution commands) and the Returner (helps the SaltStack client minion to store the execution result).

(2) Configuration management mainly records the configuration state of the target node minion by writing state files, so that the controlled minion node can synchronize to the corresponding configuration state. The state file is the core of the SaltStack configuration management. This file describes the state information that the target host needs to synchronize. The management node master issues an instruction to the target node Minion and runs the state.highstate module. The corresponding minion downloads and parses the corresponding state.sls, and then performs matching execution according to the corresponding instruction.

Through OpenStack realizing hardware resource virtualization and SaltStack implementing the remote sensing cloud application environment configuration, the automated deployment of custom remote sensing cloud environment is completed. The detailed process of automatic deployment of the remote sensing cloud environment is as follows:

The first step is parameter analysis. According to the remote sensing product order submitted by the tenant, the information is extracted from the order, mainly including computer parameter information and remote sensing element information.

The second step is cloud host deployment. The system analyzes the customized computer information (CPU cores, memory, and hard disk), and passes these parameters to the OpenStack in the background. The virtual computing hosts are deployed through the Nova computing component of OpenStack, including the number of virtual machines, operating system, CPU. core, memory, and virtual cloud host IP.

The third step is to configure the remote sensing environment, that is, to implement the related configuration of the virtual machine created by OpenStack through the server master of SaltStack. This part mainly includes the software installation configuration and the basic environment configuration of the cluster. The system transmits the remote sensing element information parameters (including the remote sensing software and remote sensing data processing algorithm libraries) to the background analysis function, obtains the target state.sls file, and executes the corresponding state.sls through the management node master, and the target client minion (The premise is that the client software minion is installed on the corresponding image of OpenStack in advance) will automatically complete the installation and configuration of the specified remote sensing environment.

9.4.2 Resource monitoring based on Ganglia

The resource monitoring of the remote sensing cloud environment includes server CPU usage, memory usage, hard disk usage, and network status, etc. The resource monitoring of the remote sensing cloud environment can provide the basis for system resource scheduling and allocation.

In order to effectively collect monitoring data and display the use of resources to users in real time, this study uses Ganglia monitoring technology and ECharts chart technology to provide remote sensing cloud resource monitoring services to remote sensing cloud users.

Ganglia is a scalable monitoring system designed for high-performance distributed systems (cluster, grid, cloud computing, etc.). The Ganglia monitoring system is based on a layered architecture that supports clusters of up to 2000 nodes and supports most operating systems such as Windows, Linux and Solaris.

Ganglia essentially monitors the running state of the cluster through a multicast-based monitoring/distribution protocol, consisting mainly of Gmond (Ganglia Monitoring Daemon) and Gmetad (Ganglia Meta Daemon).

The architecture of the Ganglia system is shown in Figure 9.8.

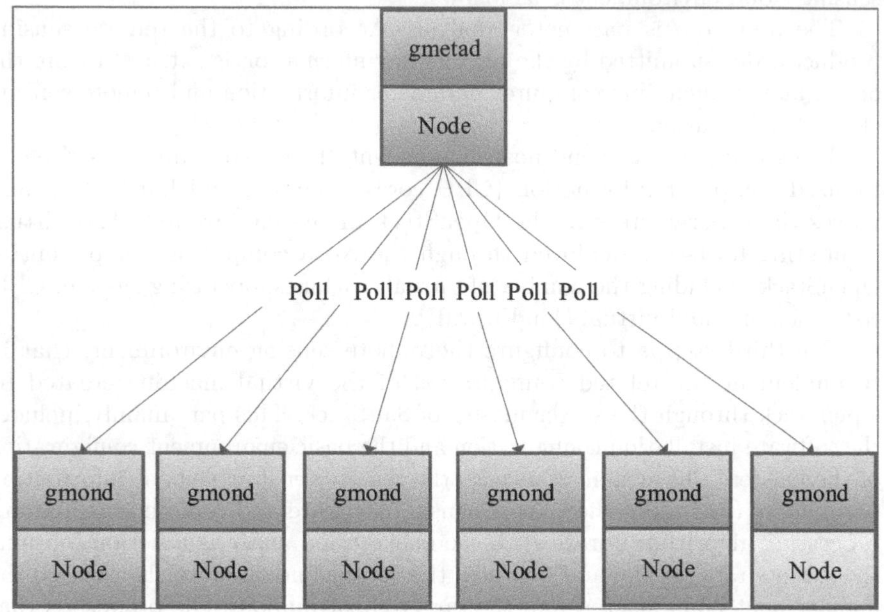

FIGURE 9.8: The architecture of the Ganglia system.

Gmond is a multicast daemon whose main function is to collect data. The Gmond program is composed of multiple threads running on each monitored node and is responsible for monitoring and collecting the health of the nodes. Gmond uses the operating system to collect real-time status data of the host, such as CPU usage, memory usage, disk I/O status, and network transmission rate. Collecting data of the Ganglia cluster through unicast or multicast has achieved the purpose of monitoring. At the same time, the ganglia cluster state request is mainly described by XML.

Gmetad is responsible for collecting the node running parameters published by Gmond. Gmetad periodically polls the data resources of each child node in the subtree, parses the collected xml, stores the values, and stores the volatile data in the round-robin database (rrd database), while the status data collected by the node is transmitted to the client.

In this way, the collected data is written to the RRDtool database. Deploying a Ganglia service on each virtual machine allows real-time monitoring of virtual machine resource usage.

ECharts is a library of pure Javascript charts, abbreviated from Enterprise Charts. ECharts is a business-grade data chart. Currently ECharts reports support running on PCs and mobile devices, and are compatible with most current browsers (IE7/8/9/10/11, Chrome, Firefox, Safari, etc.). ECharts provides a more intuitive, lively, interactive, and highly customizable data visualization chart with the underlying lightweight Canvas library ZRender [287].

The specific implementation process of resource monitoring is as follows:

(1) Running Ganglia's Gmetad on the server node to collect status data of each monitored node.

(2) Installing Ganglia's Gmond in the OpenStack system image. Gmond runs in the virtual machine provided by the remote sensing cloud system to collect monitoring data of each virtual machine in real time.

(3) Using the ECharts chart to display the running status of each cloud host in real time through the Web, providing a basis for users to maintain and manage the cloud hosts.

9.5 Conclusions

Aiming at the difficulties of traditional remote sensing computing environment construction and high cost of platform construction, this study adopts open source cloud computing architecture OpenStack and cross-platform operation tool SaltStack to realize the automatic deployment and management of remote sensing oriented cloud infrastructure with high efficiency and low cost.

The technology is mainly to develop OpenStack through the Java API-OpenStack4j, log in and manage the OpenStack cloud infrastructure platform through the back-end Java code, and combine the automatic operation and maintenance tool SaltStack to generate the State module of complex remote sensing software installation. The SaltStack related command completes the "one-click" automatic deployment of remote sensing cloud infrastructure.

This study also designs and completes the real-time accurate monitoring of cloud infrastructure resources in a remote sensing environment. The technical implementation is to collect monitoring data through Ganglia, extract data from Ganglia's RRD database, and perform a Web site visual presentation of the data. In this way, the real-time and accurate monitoring of cloud platform resources is realized for multi-tenants.

Based on how to make the remote sensing cloud provide more convenient, flexible and efficient cloud infrastructure services, this study conducts in-depth research on cloud infrastructure features and services to realize the automated deployment of cloud infrastructure in the remote sensing field.

(2) Installing Ganglia's Gmetad on the OpenStack control node. Gmetad [illegible] the [illegible] data provided by the [illegible] gmond and system to collect the status data of each node in real time.

6.5 Conclusions

Based on [illegible] the [illegible] setting up [illegible] platform [illegible] Ganglia and [illegible] cloud infrastructure services, this [illegible] realize [illegible] resource [illegible] and [illegible] services [illegible] the [illegible] of deployment cloud infrastructure [illegible].

Chapter 10

Security Management in a Remote-Sensing-Oriented Cloud Computing Environment

10.1 Introduction

In recent years, while people enjoy the high efficiency and low cost brought by cloud computing, they also face the security challenges of information. With the widespread use of cloud computing services, the security trust issue of cloud computing is more prominent. In the cloud computing environment, since the users directly use and operate the software environment, the network infrastructure, and the computing platform provided by the cloud service provider, the destruction of the cloud resource and software by the attacker is much more serious than the use of the internet for sharing resources[288]. Especially, the subjective destruction of legitimate users is more serious.

Remote sensing cloud service is based on cloud computing technology, which packages remote sensing data, information products, processing algorithm technology and computing resources into measurable services[289]. Users can obtain corresponding application and service resources according to individual needs and their hobbies through the network. It mainly includes multi-source remote sensing data distribution and sharing, high-performance data processing and product production, cloud storage, and the remote sensing computing platform[219]. However, users can directly use the computing resources of the cloud service platform. For example, IaaS (Instrument as a Service), a service platform provides a shared facility that contains components and functions, such as storage, operating system and server[290]. Those are not completely isolated for users of the system. When attacked, all servers will be transparent to the attacker. When a user accesses a cloud storage service, authentication is required to ensure the correctness and legitimacy of the user identity. The subsequent access and use of the remote sensing cloud computing platform becomes a legitimate user only when the user identity is securely and correctly authenticated. But the legitimate user cannot guarantee a secure user. Therefore whether the user behavior is trustworthy is particularly important. The assessment of user behavior credibility is an essential research content to ensure the security of the remote sensing cloud service.

In the field of cloud computing applications, many scholars have studied the application of user behavior trust evaluation methods in cloud computing environments. For example, Ji and Xianna proposed a game control mechanism for behavioral trust prediction. Firstly, a Bayesian network is used to predict the user's behavioral trust. Then, based on the combination of prediction results and game analysis, a Nash equilibrium strategy is derived[291]. Zou, Bingyu and Zhang, Huanguo and Guo, Xi and Ying, H. U. and Jamila-Sattar proposed a method to evaluate trust levels based on user behavior and platform environment characteristics [292]. Jian, X. U. and Ming-Jie, L. I. and Zhou, Fu Cai and Xue proposed an identity authentication method based on user mouse behavior, which uses a hierarchical classification decision model to authenticate user identity [293]. However, the development of this authentication mechanism does not address a specific cloud computing service. These models partially solve the problems of user trust and evaluation in different application backgrounds. However, the training set extraction of each model and method is targeted. These models lack a flexible trust evaluation mechanism, which cannot meet the personalized characteristics of user behavior evaluation in different fields.

Above all, the existing user behavior authentication methods are service-specific. User behavior of each service platform has its own characteristics. Based on traditional identity authentication and the behavior authentication mechanism, this study proposes a user behavior authentication method based on a Bayesian network model. We also established a user authentication set according to the user behavior characteristics of the remote sensing cloud service and constructed a Bayesian network model for user behavior authentication. This method mainly includes a behavior authentication process and behavior prediction algorithm. The contributions of this study are: 1) In the entire authentication process, the specific strategy is given for different situations. For users who pass the identity authentication, behavior trust prediction is performed for each behavior request. If the trusted scope is exceeded, the user's behavior request operation is rejected. The security of the certification is guaranteed in the whole certification process. 2) In terms of the authentication algorithm, the behavior attribute weight information is combined with the Bayesian network algorithm to avoid the subjectivity and uncertainty of the algorithm. It guarantees the security of the platform, and can more accurately predict the malicious behavior users of the remote sensing cloud platform, and finally realize the security of the entire authentication process.

The remainder of this chapter is organized as follows. Section 10.2 introduces the user behavior authentication scheme. Section 10.3 describes the method for user behavior trust level prediction. Finally, in Section 10.4, we conclude the study.

10.2 User Behavior Authentication Scheme

10.2.1 User behavior authentication set

The user behavior authentication set is personalized and critical to ensuring the accuracy of behavioral authentication. Therefore, the behavior authentication set is one of the most important aspects of the behavior authentication process [294, 295, 296]. In the authentication process, real-time user behavior evidence needs to be compared with the user behavior authentication set. The probability of successful user behavior authentication depends on the division, definition and behavior-related set coverage of the behavior authentication set.

The user behavior authentication set includes the following aspects: behavior state authentication set (SA), behavior content authentication set (CA), and behavior custom authentication set (HA). The behavior state authentication set is composed of behavior attributes that cause the behavior state to be abnormal, such as the user login location, the client IP address, the operating system version, and the behavior state abnormally caused by the sudden inconsistency with the historical behavior state. The behavioral state anomaly check of the PaaS layer and the SaaS layer needs to compare the operating system, such as the IP address and access time. The IaaS layer only provides infrastructure services, so there is no need to use the evidence attribute of the client's operating system [297, 298]. Certification. The IaaS layer only provides infrastructure services, so there is no need to authenticate the client's operating system property. Behavioral content mainly includes the use of user resources, such as the type and quantity of resources used. In different service modes of the cloud computing environment, the content of user access resources is different. For example, in the IaaS layer, it mainly refers to basic computing resources such as processing, storage, and network; in the PaaS layer, it refers to the development environment such as servers, operating systems, and middleware. Under normal circumstances, the number and types of resources used by users will not change greatly. If there is a large abnormality, the user's behavior may be abnormal and behavior authentication is required [299, 300]. Especially in the SaaS service model, the specific behaviors of different users are different [301]. Behavioral habits mainly include websites that users frequently visit, pages that are accustomed to entering resources, and resources that are used to access.

10.2.2 User behavior authentication process

User behavior authentication includes identity authentication and behavior authentication. According to the personalized behavior characteristics of remote sensing cloud computing platform users, a more granular behavior trust authentication scheme is designed. The design idea of the behavior authentication is as follows: User behavior data information is extracted in conjunction with server logs and client data. It includes data that has been browsed and ordered by users obtained through web logs, service information, and browsing behavior records of users entering the remote sensing cloud platform through server logs. When submitting a system access request, the user needs to perform identity authentication first. If the identity authentication fails, the user is directly refused to be provided the service. If the user identity authentication succeeds, the user enters the behavior authentication stage. Behavior authentication refers to the process of the server obtaining the real-time behavior evidence of the user and verifying the behavior evidence with the behavior authentication set stored in the database [302, 303]. In the behavior authentication process, a specific behavioral authentication set is located based on behavioral evidence. Then, the Bayesian network is used to calculate the trust level of each behavior authentication set in combination with the real-time behavior evidence. Finally, the user behavior trust level is calculated according to the decision algorithm. The specific process of the user behavior authentication scheme of the remote sensing cloud computing platform is shown in Figure 10.1. The detailed steps are as follows:

(1) When the user sends a service request to the server, the behavior authentication mechanism first performs identity authentication. For the user whose identity authentication is successful and who is not a first time log in, the user is allowed to access and implement real-time behavior monitoring, and step (2) is performed. However, for users who have successfully authenticated the identity and who are first-time log-ins to the cloud computing platform, the system allows authorized access, performs key behavior monitoring, and performs step (4); Users who fail authentication, are denied the service platform access.

(2) The user behavior status information is obtained, and then the authentication based on the behavior status authentication set is performed. For the user whose status authentication is unsuccessful, the access is denied; for the

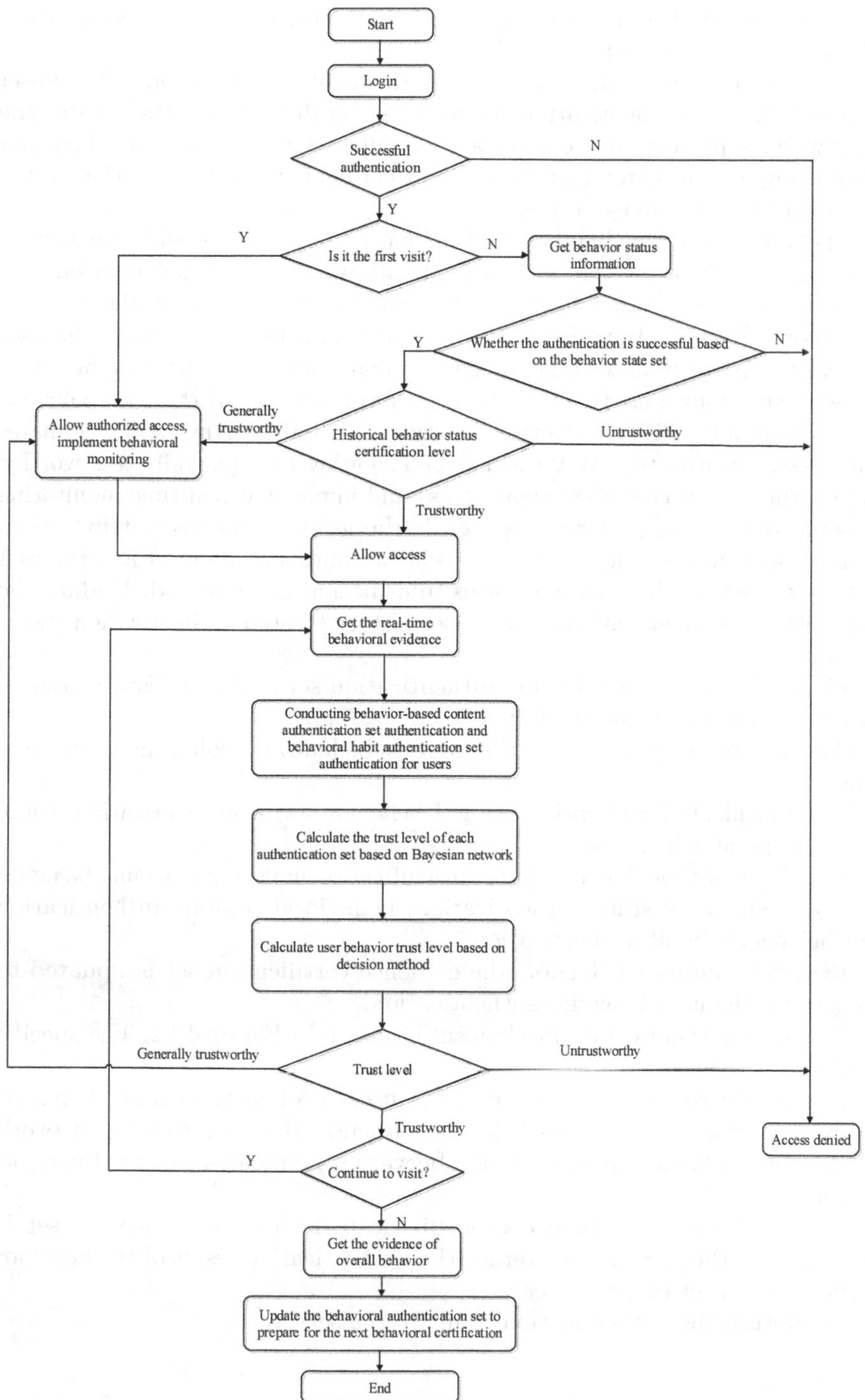

FIGURE 10.1: User behavior authentication flowchart.

user authenticated by the state authentication set, the authorized access is allowed, and step (3) is performed.

(3) Search for user history behavior authentication information. If the user history behavior authentication status is "generally trustworthy", enter the early warning prevention and allow access, but perform key real-time behavior monitoring, perform step (4); if the historical behavior authentication status is "untrustworthy", deny access.

(4) Obtain the real-time behavior evidence of the user, realize the authentication of the behavior content authentication set and the behavior custom authentication set, calculate the trust level of each behavior authentication set according to the Bayesian network model, and finally calculate the user behavior trust level according to the decision method. If the user behavior authentication level is "trusted", then perform step (2). If the user behavior authentication level is "trustworthy" or "generally trustworthy" and terminates the access, perform step (5). If the user behavior level is "generally trustworthy" and continues to access, then allow access and implement real-time monitoring of behavioral focus, perform step (2). If the level is "untrustworthy", then terminal user access. Finally, the user behavior authentication set and the user history trusted authentication status information are updated. Update the user behavior authentication set and user history trusted authentication status information.

(5) Update the user behavior authentication set and user history trusted authentication status information.

The user behavior authentication process includes the following three main processes:

(1) User identity authentication and behavior status authentication before the user operates behavior.

(2) The Real-time dynamic state authentication in user operational behavior includes behavioral state authentication, behavioral content authentication and behavioral habit authentication.

(3) After the user behavior, the evidence certification set is updated to prepare for the next behavior authentication.

The user authentication mechanism is shown in Figure 10.2. The specific steps are as follows:

(1) Obtain real-time behavior of the user through the client. Transmit behavioral evidence to the user behavior authentication server via the network.

(2) User Behavior Authentication Server authenticates captured user behavior.

(3) The behavior authentication of the behavior authentication set is performed by the server. Then return the authentication result of the behavior authentication set to the server.

(4) Return the authentication result to the server.

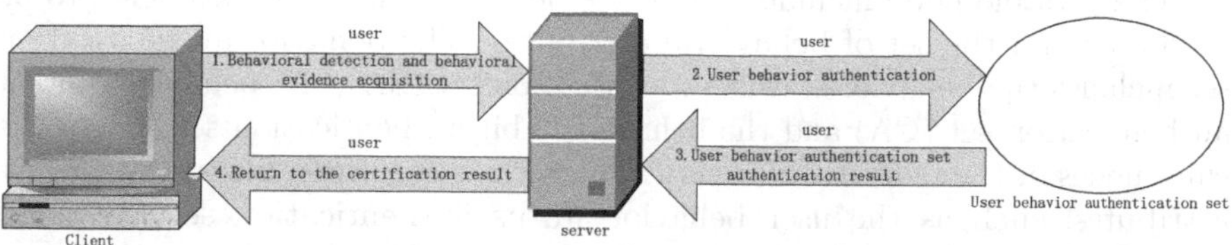

FIGURE 10.2: User behavior authentication mechanism.

The user behavior authentication process includes submitting real-time behavioral evidence obtained by the user and the system in the interaction to the corresponding server [304]. Then, the server compares the submitted behavioral evidence with the user behavior authentication set stored in the database [305, 306]. It confirms whether the user behavior is trusted according to the authentication result. If the authentication result is credible, the information service is provided to the user; otherwise the information service is refused.

10.3 The Method for User Behavior Trust Level Prediction

10.3.1 Bayesian network model for user behavior trust prediction

User behavior trust prediction is based on the user's historical interaction behavior evidence. It combines with the user's current real-time behavior to predict the user's trust level. The Bayesian network is a directed acyclic graph consisting of nodes representing variables and directed edges connecting these nodes [307]. The Bayesian network model that constitutes the prediction of user behavior trust level is shown in Figure 10.3.

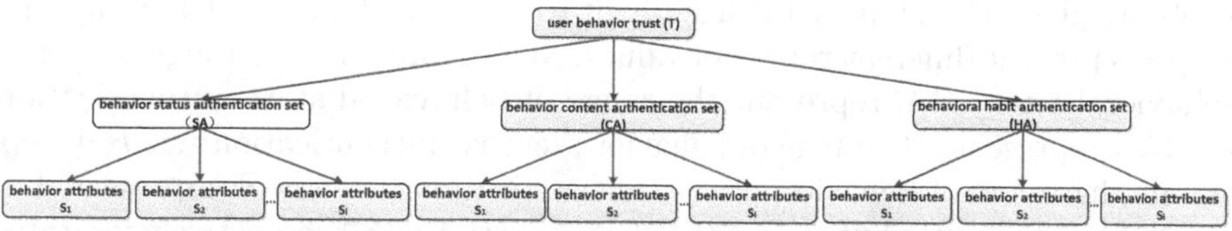

FIGURE 10.3: Basic model of Bayesian network for user behavior trust prediction.

The variable node includes the user's behavioral authentication set to be predicted and the set of behavioral attributes. The behavior authentication set includes the behavior status authentication set (SA), the behavior content authentication set (CA) and the behavior habit authentication set (HA). The child nodes of the behavior authentication set are corresponding user behavior attributes, such as the user behavior status authentication set (SA) and the behavior attributes it contains: client information, IP information, login information, etc. The user behavior content authentication set (CA) and its behavior attributes include the type of remote sensing data processing software, the number of remote sensing data downloaded, the type of remote sensing product ordered and the type of remote sensing virtual computing environment (agricultural topics, forestry topics, mineral topics, marine topics, etc.). The user behavior custom authentication set (HA) and its behavior attributes include: frequently accessed remote sensing cloud service types, frequently downloaded remote sensing data or product types, frequently ordered remote sensing virtual computing environment types, page references, and so on. The Bayesian network can visualize the user behavior trust level prediction and user behavior attributes with a directed graph. It incorporates user history and real-time behavioral statistics into the model in the form of conditional probabilities, which seamlessly combines a priori knowledge of user behavior with a posteriori data. Moreover, the nodes in the Bayesian network interact with each other. The change of the value of any node affects other nodes, thus meeting the reasoning and prediction effects of fine-grained combinations of different requirements.

10.3.2 The calculation method of user behavior prediction

10.3.2.1 Prior probability calculation of user behavior attribute level

When using a Bayesian network for user behavior authentication, the prior probability of user behavior attributes needs to be calculated. In this chapter, the user behavior trust (T), the behavior status (SA), the behavior content (CA), the behavior habit (HA), and the behavior attribute corresponding to each behavior authentication set are divided into L trust levels, and the trust levels are given the number i (i from high to low. =1, 2, 3.. L). The array subscripts represent different ranges of values, so Ti represents the range of overall behavioral trust, SA_i represent the range of behavioral state authentication set, CA_i represents the range of behavior security authentication set, HA_i represents the range of behavior custom authentication set, and Si represents the range of behavior attributes. $|T_i|, |SA_i|, |CA_i|, |HA_i|, |S_i|$ respectively indicates the number of times the value of the overall trust, the behavior authentication set, and the behavior attribute of the user interaction history falls within the range of T_i, SA_i, CA_i, HA_i, and S_i, respectively. The "n" represents the total number of user interactions. $P(T_i), P(SA_i), P(CA_i), P(HA_i), P(S_i)$ represent their probabilities, respectively. The meaning of these symbols is applicable throughout the chapter.

The prior probability calculation formula for the user behavior authentication set level is as follows:

$$P(T_i) = \frac{|T_i|}{n}, \qquad (1 \leq i \leq 3, \sum_{i=1}^{3} P(T_i) = 1) \tag{10.1}$$

The method of calculating the prior probability of the user behavior attribute set level is similar to the method of calculating the prior probability of the user behavior authentication set level. The prior probability calculation formula of the behavior attribute trust level is as follows:

$$P(S_i) = \frac{|S_i|}{n}, \qquad (1 \leq i \leq L, \sum_{i=1}^{L} P(S_i) = 1) \tag{10.2}$$

10.3.2.2 Conditional probability of behavioral authentication set

In addition to calculating the prior probability, the conditional probability of each behavioral certification set level is also required. The conditional probability formula for the certification set level is as follows:

$$P(e/h) = \frac{p(h,e)}{p(h)} \tag{10.3}$$

It indicates the probability of satisfying the e under the condition that h is satisfied. Taking the conditional probability of $p(S_i|CA_j)$ as an example, it indicates the probability that the behavior attribute node falls within the S_i range under the condition that the behavior content authentication level is j. The calculation formula is as follows:

$$P(S_i|CA_j) = \frac{p(S_i, CA_j)}{P(CA_j)} = \frac{S_i ICA_j/n}{|CA_j|/n} = \frac{S_i ICA_j}{|CA_j|} \tag{10.4}$$

The probability of the behavior authentication set trust level can be calculated from the a priori probability and the conditional probability of each behavior attribute set level obtained by the above calculation method. In the following, under the single-behavior attribute condition, taking the predictive behavior content authentication set trust level as an example, the other behavior authentication set calculation method is similar. Then, we calculate the probability $P(CA_i|S_j)$ of the behavior content certification level with the behavior attribute attribute level falling within the S_j range. Combined with the Bayesian formula, the formula is as follows:

$$P(CA_i|S_j) = \frac{P(S_j|CA_i)P(CA_i)}{P(S_i)} \tag{10.5}$$
$$= \frac{(|S_jICA_i|/|CA_i|)(|CA_i|/n)}{|S_i|/n}$$
$$= \frac{|C_jICA_i|}{|S_j|}$$

For multiple behavior attributes, other behavior authentication set trust level predictions are similar to the predicted behavior content authentication sets. The following is an example of predicting behavior content authentication sets. Assume that there is behavior content. The authentication set contains four behavior attributes: S1, S2, S3, and S4, which fall under the conditions of $S1_j$, $S2_k$, $S3_m$, $S4_n$, and the behavior authentication set CA trust level is predicted. The calculation formula is as follows:

$$P(CA_i|S1_j, S2_k, S3_m, S4_n) = \frac{P(S1_j, S2_k, S3_m, S4_n|CA_i)P(CA_i)}{P(S1_j, S2_k, S3_m, S4_n)} \tag{10.6}$$
$$= \frac{(|S1_jIS2_kIS3_mIS4_n|/|CA_i|)(|CA_i|/n)}{|S1_jIS2_kIS3_mIS4_n|/n}$$
$$= \frac{|S1_jIS2_kIS3_mIS4_nICA_i|}{|S1_jIS2_kIS3_mIS4_n|}$$

10.3.2.3 Method of calculating behavioral trust level

After calculating the trust level of each behavior authentication set, the weight information of each behavior authentication set is analyzed according to the graphical Bayesian network. Finally, the user behavior trust level T is calculated according to the polynomial calculation method:

$$T = \sum_{i=1}^{i} C_i W_i \qquad (W_i) \geq 0, \sum_{i=1}^{i} W_i = 1) \tag{10.7}$$

In the above formula, C_i represents the trust level of the i-th behavior authentication set, and W_i represents the weight of the i-th behavior authentication set.

10.3.3 User behavior trust level prediction example and analysis

According to the user behavior trust level prediction method proposed above, an example will be used to demonstrate the validity of the prediction model. At some point, user(U) requests access to server(S), which now predicts the behavior trust level of user(U) under certain conditions of the behavior attribute set. The experiment is based on 207 sets of data of user interaction with the remote cloud service platform. The behavior trust is divided into three levels: trustworthy, generally trustworthy, and untrustworthy. The user's behavior content trust level prediction process is given below. The frequency of the user behavior evidence falling in different behavior attribute intervals (S1, S2, and S3 represent three different behavior attributes) is shown in Table 10.1.

TABLE 10.1: The frequency of user behavior evidence falls within the behavior attribute set interval.

scope	Frequency of S1	scope	Frequency of S2	scope	Frequency of S3
S11	33	S21	47	S31	36
S12	73	S22	98	S32	89
S13	101	S23	62	S33	82

(1) The calculation of the trust level of the behavior authentication set.

According to the user behavior interaction data information obtained above, the prior probability of each node in the network can be calculated by using formula (10.2). The prior probability of the behavior attribute node is shown in Table 10.2. The conditional probability table of the behavior authentication level node T is shown in Table 10.3. Then, the conditional probability of each evidence attribute is calculated by formula (10.4), and the conditional probability table CPT of each evidence attribute is calculated. The prior probability of all nodes and the conditional probability of the child nodes are calculated, which ensures that there is a corresponding prior probability and conditional probability for any set of observations when calculating the trust level prediction of the behavior authentication set. Then, by bringing them into equation (10.5), you can calculate all posterior probabilities. Finally, the maximum probability value is taken as the trust level of each behavior authentication set.

TABLE 10.2: The prior probability of behavior attribute nodes.

Node	Prior probability	Node	Prior probability	Node	Prior probability
S11	33/207	S21	47/207	S31	36/207
S12	73/207	S22	98/207	S32	89/207
S13	101/207	S23	62/207	S33	82/207

(2) The calculation of user behavior trust level

According to the prediction step of the user behavior trust level mentioned above, the Bayesian network model is used to analyze the Bayesian network structure relationship diagram of each behavior attribute. The result is shown in Figure 10.4. From this Bayesian network structure diagram, we can analyze

TABLE 10.3: The prior probability of behavior authentication node.

T	P(T)
1	1/3
2	1/3
3	1/3

the weight information of each behavior attribute to predict the trust level of the user behavior. In this experiment, the user behavior attribute weight distribution map is shown in Figure 10.5. The weight values of each behavior attribute are: page reference weight is 0.225, browsing address URL weight is 0.190, access time weight is 0.190, and byte weight is 0.165, the resource weight is 0.225, and the client information weight is 0.072. According to the weight value of the behavior attribute, the weighted average algorithm is used to calculate the weight value of each behavior authentication set. The weight information of the behavior state authentication set obtained by the experiment is as follows: behavior state authentication set weight W_s=0.392, behavior content authentication set weight W_c=0.402, behavior habit authentication set weight W_h=0.206.

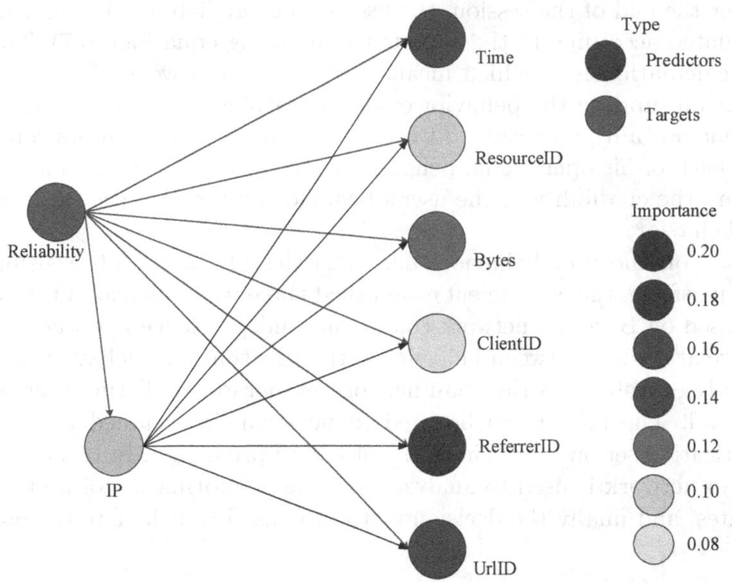

FIGURE 10.4: Bayesian network.

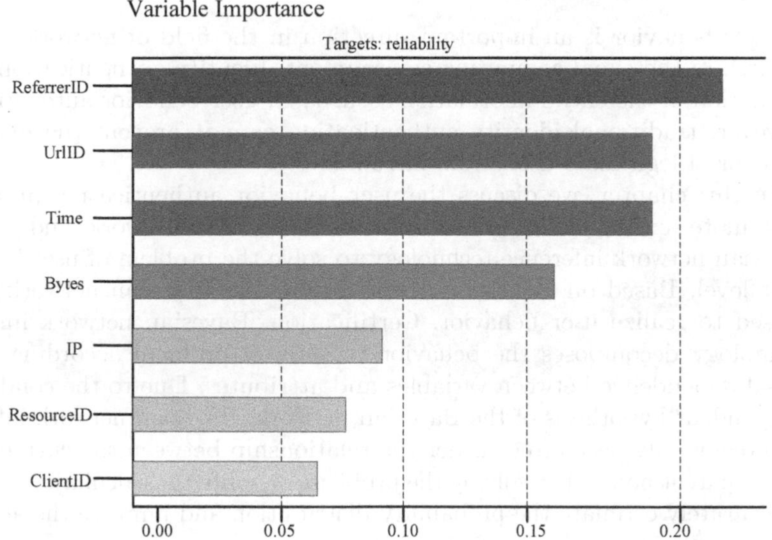

FIGURE 10.5: Variable importance.

After the end of the session, the user's final predicted behavior trust level is calculated according to the weighted value using equation(10.7). The result of the calculation is 2, which means "generally trustworthy". After ending this session, update the behavior evidence database to prepare for the next user behavior authentication. At the same time, when the users continue to access, each of his operational behavior requests needs to be re-authenticated to ensure the credibility of the user's behavior and ensure the security of the user's behavior.

Based on the user behavior characteristics of the remote sensing cloud service platform, the experiment established three user behavior authentication sets. Based on Bayesian network conditional independence hypothesis and the causal relationship between behavior authentication and behavior attributes, this study establishes a Bayesian network model for predicting user behavior trust level. The model can be used to perform fine-grained user behavior authentication set on user behavior. Trust level prediction. In this chapter, the Bayesian network is used to analyze the weight information of user behavior attributes, and finally the decision method is used to calculate the user trust level.

10.4 Conclusions

User behavior is an important direction in the field of network security research in the cloud computing environment. Identity authentication is the foundation of the entire information security in user behavior authentication. However, traditional identity authentication cannot prevent the malicious behavior of legitimate users from intruding.

In this chapter, we discuss the user behavior authentication process of the remote sensing cloud service based on a Bayesian network, and combine Bayesian network inference technology to solve the problem of user behavior trust level. Based on identity authentication, the Bayesian network model is used to realize user behavior. Certification. Bayesian network inference technology decomposes the behavior trust level problem according to the causal dependence between variables and attributes. Due to the conditional independent hypothesis of the Bayesian network, Bayesian network inference technology only needs to consider the relationship between the current node and its parent node when solving the problem, simplify the calculation and help to accurately estimate the probability distribution and improve the accuracy of the prediction.

Finally, such programs will face many attacks in the future. Due to the lack of experience in research and use, the challenges they face and the viability of defensive attacks require further research.

Chapter 11

A Cloud-Based Remote Sensing Information Service System Design and Implementation

11.1 Introduction

The development of Earth Observation technology has increased data volume growth, shortened data acquisition and update cycles, and improved data timeliness. On the one hand, it has led to an increase in the variety and quantity of remote sensing products, and expanded the application fields of remote sensing. On the other hand, it has brought great challenges to the data storage and management, data processing and products production, and proposed more efficient, faster, and more convenient requirements for remote sensing information services [308, 309].

For remote sensing data storage and management, the differences in storage format, projection standard, resolution, and revisit cycle of different types of remote sensing data sources have resulted in the difficulties in data integration, data organization, rapid data retrieval and access. For multi-source remote sensing data processing and product production, the production process is fixed and needs to be prepared in a manual manner in advance, in the existing remote sensing data production systems [310]. It is impossible to organize the production process according to the individual needs of users, and it is unable to meet the user's individual, complex and varied production needs. For remote sensing data users, they need to download large amounts of remote sensing data from USGS, NOAA, NASA and ESA websites, and use their own workstations to store and process such big data. It would be challenging for such users to obtain the remote sensing information they need [189].

In view of the above mentioned factors, many national and international infrastructure projects have been conducted or are on-going, including Global Earth Observation System of Systems (GEOSS) [311], the European Commissions INSPIRE [312], and the U.S. NSF EarthCube [313]. These information infrastructures have been benefiting scientific communities and supporting scientific research. Therefore, it is one of the key research directions to promote the construction of spatial information infrastructure and combine the current new technologies and methods in the computer field to improve the level of information services [314, 315].

Cloud computing, which is a new distributed computing model after super-computing, cluster computing and grid computing [316, 317], not only has the capabilities of distributed massive data storage, High-Performance Computing (HPC), virtualization, flexible expansion, and on-demand services [318, 319], but also provides users with more personalized, highly open, flexible and convenient high-efficiency services. Using cloud computing technology in the field of remote sensing cannot only improve the efficiency of massive remote sensing data storage management, processing and product production, but also realize resource sharing and on-demand service modes, thereby improving the timeliness of remote sensing information services [320, 321].

Therefore, based on the advantages of cloud computing's massive data storage, HPC, elastic expansion, on-demand services, etc., through the research on large-scale multi-source remote sensing data integration and management, high-performance processing and product production, and automatic construction of remote sensing virtual computing environment, we designed and developed the Cloud-based Remote Sensing Information Service (CRSIS) system. The proposed CRSIS system provides four types of user-oriented [322] cloud services, namely, 1) Remote Sensing Data as a Service (RSDaaS); 2) Remote Sensing Data Processing as a Service (RSDPaaS); 3) Remote Sensing Product Production as a Service (RSPPaaS); and 4) Remote Sensing Computing Platform as a Service (RSCPaaS). Thereby, it would provide industrial application solutions for ecological environment monitoring [323], land resources survey [324], and smart cities [325, 79].

The chapter is organized as follows. Section 11.2 outlines its information service mode. Section 11.3 describes the overall architecture of the CRSIS system, and Section 11.4 details the function module of the proposed cloud service system. Section 11.5 presents a prototype system, and Section 11.6 summarizes the conclusion and provides recommendations for further research.

11.2 Remote Sensing Information Service Mode Design

11.2.1 Overall process of remote sensing information service mode

The cloud-based remote sensing information service, aims to meet the needs of users and provides four types of information services, including remote sensing data service, remote sensing data online processing service, remote sensing products production service and remote sensing computing platform service. In addition, in order to provide users with convenient data storage containers, the cloud storage service is also prepared. The overall process of information service mode is shown in Figure 11.1.

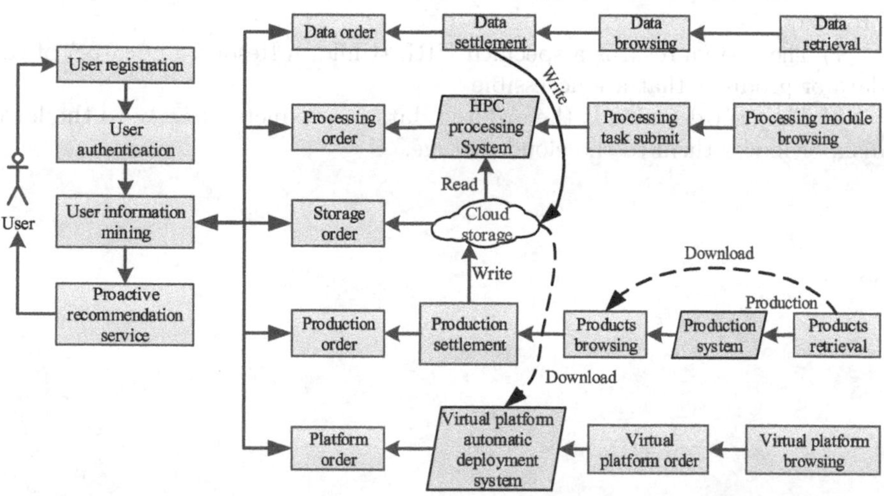

FIGURE 11.1: The overall flow chart of remote sensing information service.

11.2.2 Service mode design of RSDaaS

Remote Sensing data service includes original remote sensing data services and remote sensing value-added products [326] services, two types. It mainly provides users with remote sensing original data and value-added product retrieval, browsing, ordering, settlement, downloading and moving cloud storage services [327].

The UML(Unified Modeling Language) sequence diagram of the remote sensing data service is shown in Figure 11.2, and the detailed steps are as follows.

1) The user inputs original data or product retrieval parameters through the data service interface so as to get what he or she wanted.

2) The data service system executes the query and returns the retrieval results according to the users' retrieval parameters.

3) For the original remote sensing data service, user can directly browse the returned query results and select the most suitable remote sensing data; and for the remote sensing value-added product service, if the user intended value-added products do not exist, he or she needs to enter the product production system and submit production order, then obtain what he or she wanted of remote sensing value-added products.

4) The user adds the remote sensing data or products that he or she considers to be the most appropriate or intended to the shopping cart, waiting for settlement.

5) The user orders the remote sensing data or products in the shopping cart.

6) The user pays for what he or she ordered in remote sensing data or products.

7) The system returns a specified URL (Uniform Resource Locator) of the data or products that are accessible.

8) The user downloads the ordered data or products directly to the local disk, or moves them to the cloud storage.

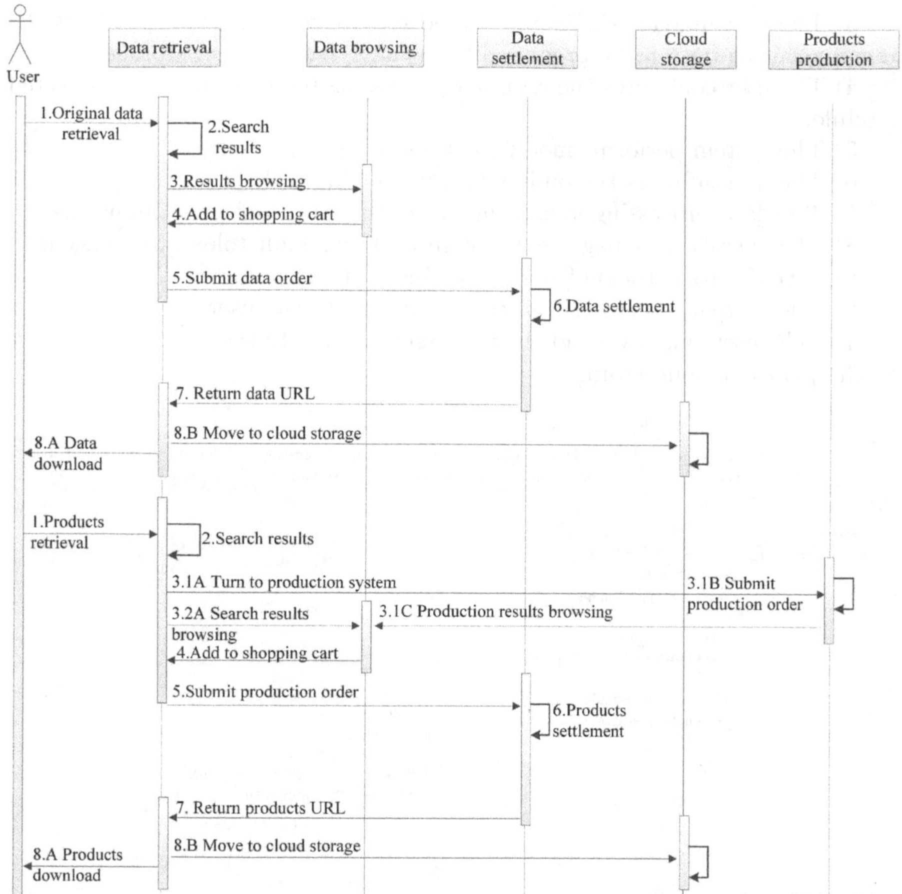

FIGURE 11.2: The UML sequence diagram of remote sensing data service.

11.2.3 Service mode design of RSDPaaS

Remote sensing data online processing service, based on virtual HPC clusters, mainly provides users with online image enhancement, transformation, mosaic, fusion, information extraction and other services [328]. The UML sequence diagram of the remote sensing data online processing service is shown in Figure 11.3, and the detailed steps are as follows.

1) The user browses the list of data processing modules by registering and logging into the data processing service portal.

2) The user selects the required data processing module according to the type and volume of data that he or she needs to process.

3) The user uploads the locally stored remote sensing data, or selects the remote sensing data to be processed from the personal cloud storage.

4) The user configures the required processing parameters for the selected module.

5) The system performs module parameter integrity check.

6) The user submits the online data processing task.

7) The data processing system monitors the status of the running task.

8) The operation management system performs fault-tolerant management to improve the robustness of the processing system.

9) The system returns the processing results to the user.

10) The user can download the processing results to the local disk or move to the personal cloud storage.

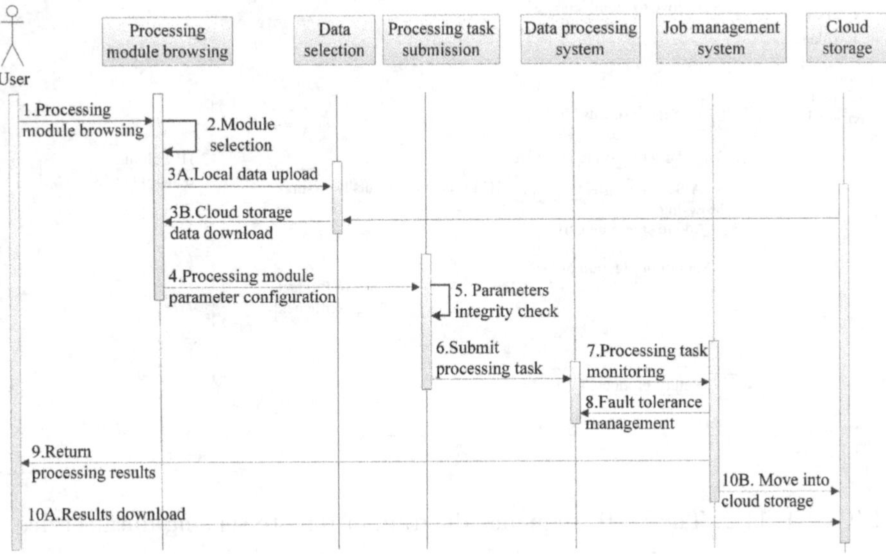

FIGURE 11.3: The UML sequence diagram of remote sensing online data processing service.

11.2.4 Service mode design of RSPPaaS

The processes of RSPPaaS mainly include production order submission, order analysis, order confirmation, production workflow organization, production workflow execution and management, etc. [329]. The UML sequence diagram of RSPPaaS is shown in Figure 11.4, and the detailed steps are as follows.

1) The user submits the production order through the service portal.

2) The product production system performs order analysis, clarifies the data, products and workflow required, and analyzes the feasibility of the user submitted production order.

3) According to the order analysis results, the RSPPaaS system carries out the required data and product inquiry.

4) The system returns data and product query results.

5) According to the data and product query results, the system will organize workflows.

6) The organized workflow checks the completeness of the data set returned in step 5). The completeness check here includes two types: time range completeness and spatial range completeness check.

7) If the data set is complete, you can directly select the workflow to perform product production; if the data set is incomplete, you need to schedule the data resources as needed, and reorganize the data-complete workflow.

8) The data-complete workflow will be submitted to the product production system.

9) The system returns the running status and result information of the production workflow, in order to be convenient for the workflow management autonomously.

10) The system returns the completed remote sensing products, and performs products registration and archiving.

11) The system returns the generated products to the user, which can be downloaded or moved into the cloud storage directly.

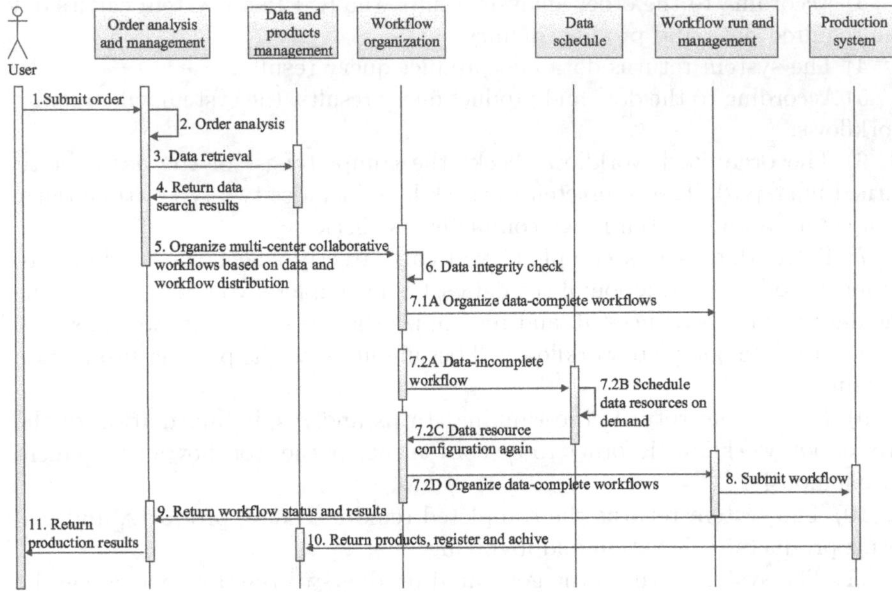

FIGURE 11.4: The UML sequence diagram of remote sensing product production service.

11.2.5 Service mode design of RSCPaaS

The RSCPaaS mainly provides users with virtual machines or clusters that integrate commonly used remote sensing data processing softwares (ENVI, etc.), algorithm libraries (GDAL libraries, etc.). The UML sequence diagram of the RSCPaaS is shown in Figure 11.5, and the detailed steps are as follows.

1) The user submits the remote sensing computing platform order according to individual needs, which mainly includes virtual machine operating system information, system basic configuration information, remote sensing data processing software information, algorithm library information, etc.

2) The RSCPaaS system performs order analysis to determine the computing resources and software resources required for platform deployment;

3) According to the order analysis result, the system queries the platform resource library for the required computing resources and software resources.

4) The system returns the query results and informs the user of the order feasibility.

5) The user pays for the ordered platform.

6) The system automatically deploys the remote sensing computing platform according to the user's order.

7) The system returns to the user with the successfully deployed remote sensing computing platform.

8) The user uses the platform and makes further new feature development based on the platform.

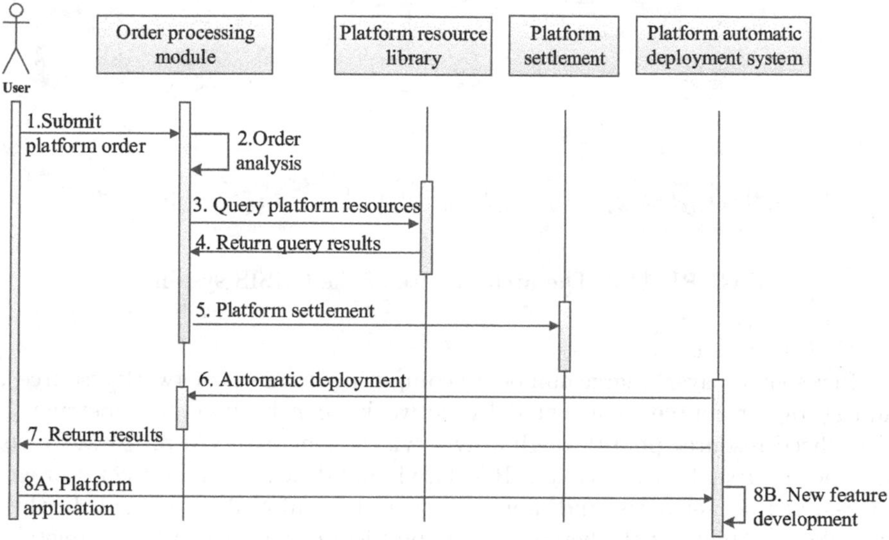

FIGURE 11.5: The UML sequence diagram of remote sensing computing platform service.

11.3 Architecture Design

The CRSIS system was developed based on the OpenStack cloud computing framework. Its system architecture can be roughly divided into five layers [330, 331], from bottom to top, namely resource layer, management layer, computing layer, business layer and service layer (Figure 11.6).

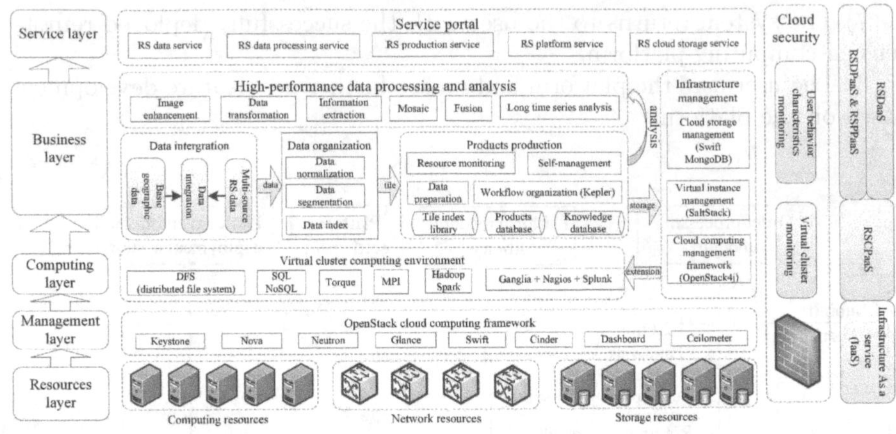

FIGURE 11.6: The architecture of the CRSIS system.

1) Resources layer

In resources layer, a large number of computing resources, network resources, and storage resources, connected by networks, can be used to construct a virtualized resource pool through a hypervisor (such as KVM or QEMU), so as to form virtual CPUs, virtual RAMs, virtual disks, virtual object storage spaces, virtual networks, and more, that can be uniformly managed. In this way, the variability of the heterogeneous physical resources can be completely shielded, and the ready-to-use logical resources can be available to users [332].

2) Management layer

The management layer mainly adopts the OpenStack computing framework and uses its core components to manage each virtual resource. The Keystone component mainly provides authentication and management services for account and role information of users in the cloud environment; the Nova component can quickly create virtual machine instances based on users' needs, and is responsible for lifecycle management of virtual machines or virtual clusters; By defining virtual networks, virtual routes, and virtual subnets, the Neutron components provide virtual network services in a cloud computing environment, including Flat, FlatDHCP, and VLAN modes; the Glance component is a virtual machine image management component that is responsible for creating, registering, querying, updating, deleting, and editing virtual machine images. It can support multiple image formats such as QCOW2, AKI, and AMI; the Swift component mainly provides virtual containers for object storage, supporting large-scale and scalable systems, with high fault tolerance and high availability features; the Cinder component mainly provides a data block storage service for a running virtual machine instances; the Horizon component provides a web-based modular user interface for OpenStack service management to simplify user operations on the service; the Ceilometer is a resource pool monitoring component, that records all resource usage of the cloud-based system, including user usage reports and user statistics, and also provides data support for system expansion and user billing.

3) Computing layer

The computing layer mainly provides virtual cluster computing environments, including remote sensing data storage containers, cluster computing and scheduling, and computing environment monitoring services.

Remote sensing images are stored in the MongoDB-based OpenStack-Swift object storage container. Each image is stored as an object in the Swift object storage container, and each object is given a domain name address, which can be used to be accessed by the user. MongoDB mainly realizes distributed storage and management of metadata and catalog data of remote sensing images, and provides fast metadata retrieval capability [333]. The virtual machine instances of the virtual cluster are mainly scheduled and generated by the OpenStack-Nova component. The storage space of the virtual computing cluster is provided by the block storage component OpenStack-Cinder. In addition, automatic deployment and system expansion of the virtual remote sensing computing environment can be realized based on the SaltStack [286] management tool.

Computational environment monitoring includes Ganglia-based resource monitoring, Splunk-based log monitoring and mining, and Nagios-based resource anomaly alarm, three parts. Ganglia can provide real-time static monitoring data and system performance metrics for the computing systems, such as platform network service monitoring, node resource monitoring (including CPU usage, memory usage, disk usage, etc.), and computing task monitoring. The Syslogd server is responsible for recording the system logs of the distributed computing and storage nodes, and periodically sends the node log information to the log collection server syslog-ng. Subsequent log information summary and classification cleaning, as well as generating log reports and early warning analysis are also completed by Syslogd. The abnormal resource alarm refers to a system that is overloaded with system resources, causing a sharp drop in system performance or even causing the system to crash, thus issuing a warning. In this case, the system can adjust the virtual machine resources according to the alarm situation, and the system administrator can conduct investigation and analysis of the system operation status [334].

4) Business layer

The business layer is the core of the entire cloud service system, mainly providing remote sensing data integration and management, data organization, remote sensing product production, high-performance data processing and analysis [335], remote sensing computing platform automatic deployment, remote sensing cloud storage and other services.

The businesses of remote sensing data integration and management, data organization, and product production run in a multi-center virtual cluster, which is composed of a main center and multiple sub-centers (Figure 11.7). In the data integration and management process, each sub-center is responsible for storing different types of remote sensing data, and the main center uses the data crawler to regularly ingest the remote sensing image metadata and thumbnails of each sub-center to achieve unified integration and management; the data organization process mainly performs geometric precision correction and fine radiation correction, image segmentation and data tile index building, so as to provide an Analysis Ready Data (ARD) [336] set for further data processing and product production; in the production process, the main center is responsible for accepting and parsing the product production orders submitted by the user, then determine which sub-centers the required data is stored in, and then distribute product production tasks to that sub-center according to the principle of 'minimum data migration', finally summarize product production results and provide feedback to users. The principle of 'minimum data migration', that is, in a product production task, is that if a sub-center A can provide the largest amount of data, then the production task will be deployed in A, and the other required data stored in other sub-centers should be migrated to A.

The business of high-performance data processing and analysis is completed by virtual HPC clusters deployed with MPI [337] and Hadoop, providing image mosaic, image fusion, feature extraction and long-time analysis, etc. [338, 339].

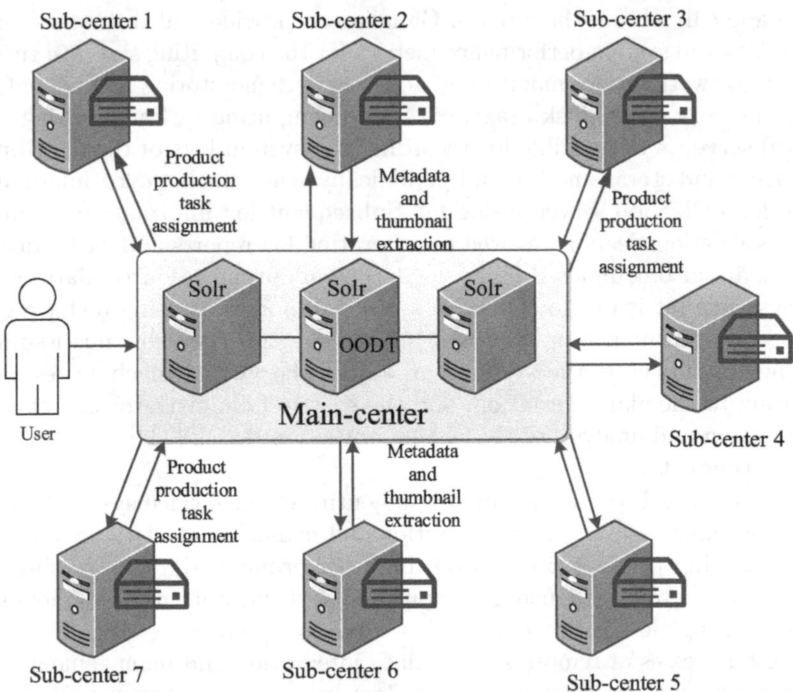

FIGURE 11.7: The framework of the virtual multi-center cluster.

5) Service layer

The service layer is a service portal of the CRSIS system. It mainly provides users with interfaces for data services, data processing services, product production services, computing platform services, and cloud storage services. In addition, in order to ensure the security of the CRSIS system, apart from deploying a firewall at the resource layer and performing virtual computing cluster monitoring at the computing layer, user behavior characteristics monitoring also needs to be performed on the service layer.

11.4 Functional Module Design

11.4.1 Function module design of RSDaaS

The RSDaaS function modules mainly include distributed multi-source remote sensing data integration system, distributed data management system, multi-user data distribution and acquisition system, cloud-storage-based data sharing system, user management system, virtual resource management system and other functional modules (Figure 11.8).

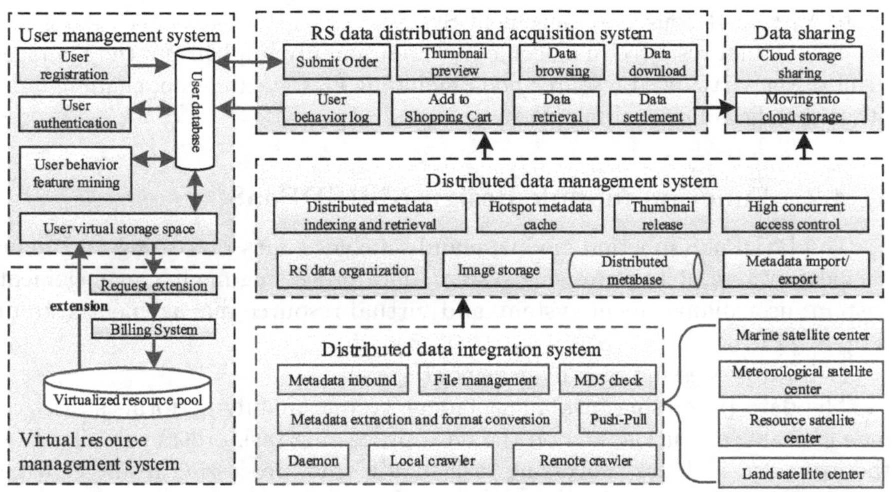

FIGURE 11.8: The data service module of the CRSIS system.

1) Distributed multi-source remote sensing data integration system

The distributed multi-source remote sensing data integration system, which was developed on the basis of the open source Object Oriented Data Technology

(OODT) [340], mainly realizes multi-source remote sensing data integration in distributed data center scenarios, including distributed data ingestion, metadata extraction and format conversion, MD5 file verification, data transmission, metadata storage, file management and other specific functional modules.

2) Distributed data management system

The distributed data management system, based on distributed data server SolrCloud, provides rapid indexing and retrieval of massive remote sensing data, including remote sensing data organization, image storage, metadata storage and distributed index construction, distributed retrieval, hotspot metadata cache, high concurrent access control, remote sensing data thumbnail release and other specific functional modules.

3) Remote sensing data distribution and acquisition system

The remote sensing data distribution and acquisition system mainly implements user-oriented data distribution and acquisition, including data retrieval, data browsing, thumbnail preview, data order, data settlement, data downloading, and other specific functional modules.

4) Remote sensing data sharing system

The remote sensing data sharing system mainly implements cloud-storage-based data sharing.

5) User Management System

The user management system mainly implements user registration, user authentication, virtual storage space management, and user behavior characteristics analysis.

6) Virtual Resource Management System

The virtual resource management system mainly implements the management of the virtualized resource pool, including elastic extension, charging and other specific functional modules.

11.4.2 Function module design of RSDPaaS

The RSDPaaS function module mainly includes data processing task management system, data processing system, data processing module management system, user management system, and virtual resource management system (Figure 11.9).

1) Data processing task management system

The data processing task management system mainly performs statistics, management, execution, etc. on the data processing task orders submitted by the user, and includes processing task monitoring, processing result viewing and downloading, processing parameter integrity checking, data preparation and scheduling and other specific function modules.

2) Data processing system

The data processing system is mainly based on virtual HPC clusters to realize rapid processing of a huge amount of remote sensing data, mainly including processing task resource matching and scheduling, processing task execution and management, etc.

3) Processing module management system

The processing module management system mainly implements the computing and storage resource management of the high-performance remote sensing data processing module, and realizes flexible expansion according to the complexity of the processing task.

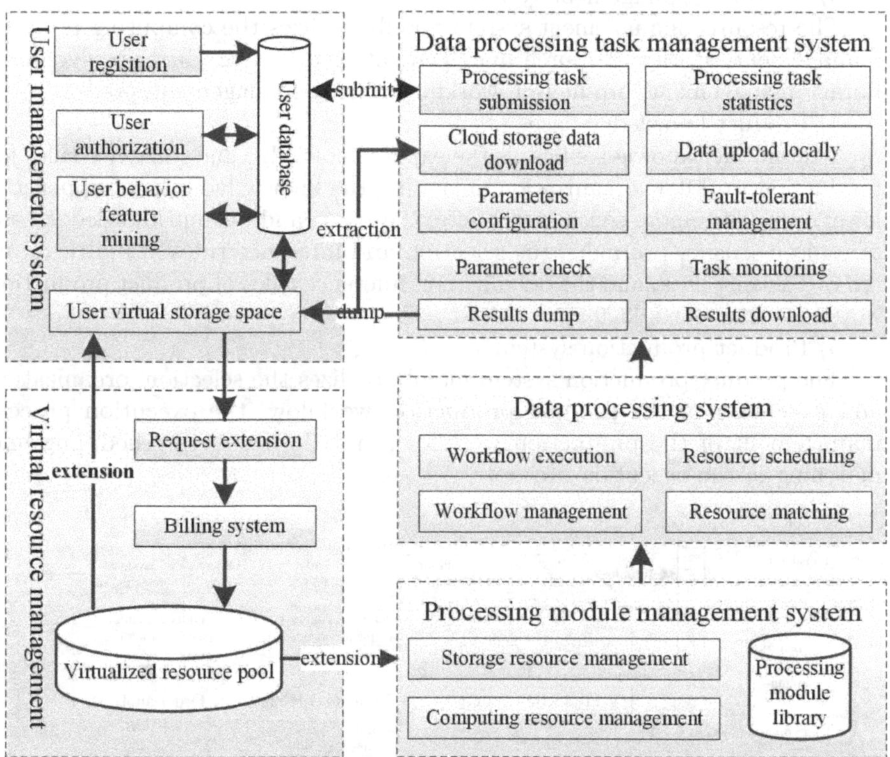

FIGURE 11.9: The data online processing service module of the CRSIS system.

11.4.3 Function module design of RSPPaaS

The RSPPaaS function module mainly includes the production order management system, order analysis system, resource management system, product knowledge base, production system, user management system, virtual resource management system and other functional modules (Figure 11.10).

1) Production order management system

The order management system mainly implements data processing and product production order receipt, statistics, execution monitoring, order log collection, order processing result viewing and downloading, and results transferring to cloud storage.

2) Order analysis system

The order analysis system mainly implements the analysis of data processing and product production orders submitted by users, including data requirement analysis, products requirement analysis, workflows requirement analysis, and feasibility determining of production orders.

3) Resource management system

The resource management system mainly realizes the computing resource management, storage resource management, product production algorithm library management, production workflow library management, etc.

4) Product knowledge base

The product knowledge base is the expert knowledge and inference rules for production workflow organization, including the knowledge base of upper and lower levels of remote sensing products, the input and output knowledge base of remote sensing products, the selection and inference rules of multi-source remote sensing data, and the organization inference rules of product production workflow, four parts [329].

5) Product production system

The product production system mainly realizes the selection, organization and execution of the product production workflow, the execution process management of the production workflow, and the resource scheduling and matching in the execution process.

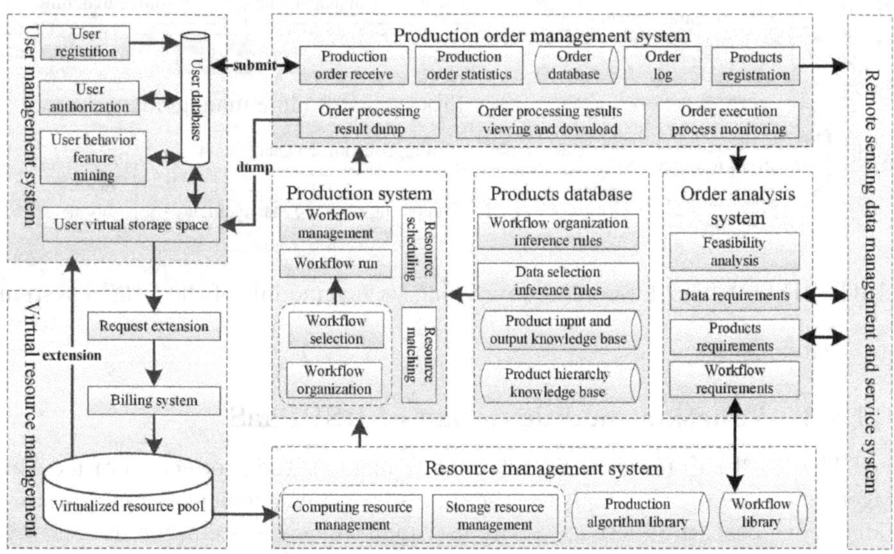

FIGURE 11.10: The product production service module of the CRSIS system.

11.4.4 Function module design of RSCPaaS

The RSCPaaS function module mainly includes the operating system, software, and library management system, computing platform ordering system, and platform automatic deployment system (Figure 11.11).

1) Operating system, software, and library management system

It mainly realizes mirror registration of the operating system image, remote sensing data processing software and remote sensing data processing library, and uses their corresponding metadata to build a database for efficient management.

2) Computing platform ordering system

It is mainly responsible for the operating system, software, and images selection, as well as platform orders generation and price accounting.

3) Computing platform automatic deployment system

The automatic deployment is mainly through the configuration of SaltStack to achieve automatic deployment of the operating system, and use the automatic deployment scripts corresponding to each software and library to realize the automatic deployment.

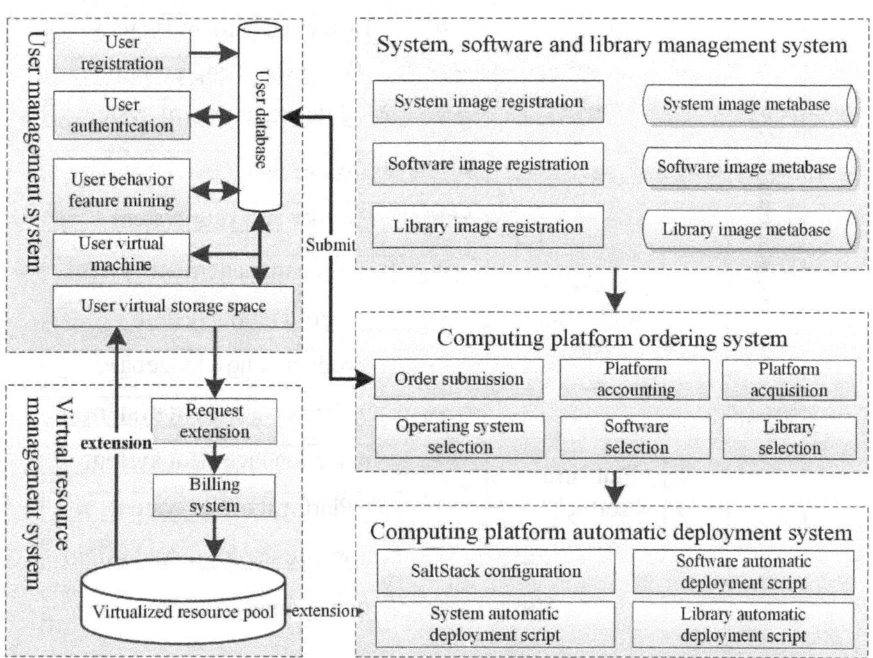

FIGURE 11.11: The platform service module of the CRSIS system.

11.5 Prototype System Design and Implementation

The prototype system of CRSIS was mainly composed of the RSDaaS subsystem, RSDPaaS subsystem, RSPPaaS subsystem, RSCPaaS subsystem and cloud storage system on the virtualized resource pool constructed by OpenStack (Figure 11.12).

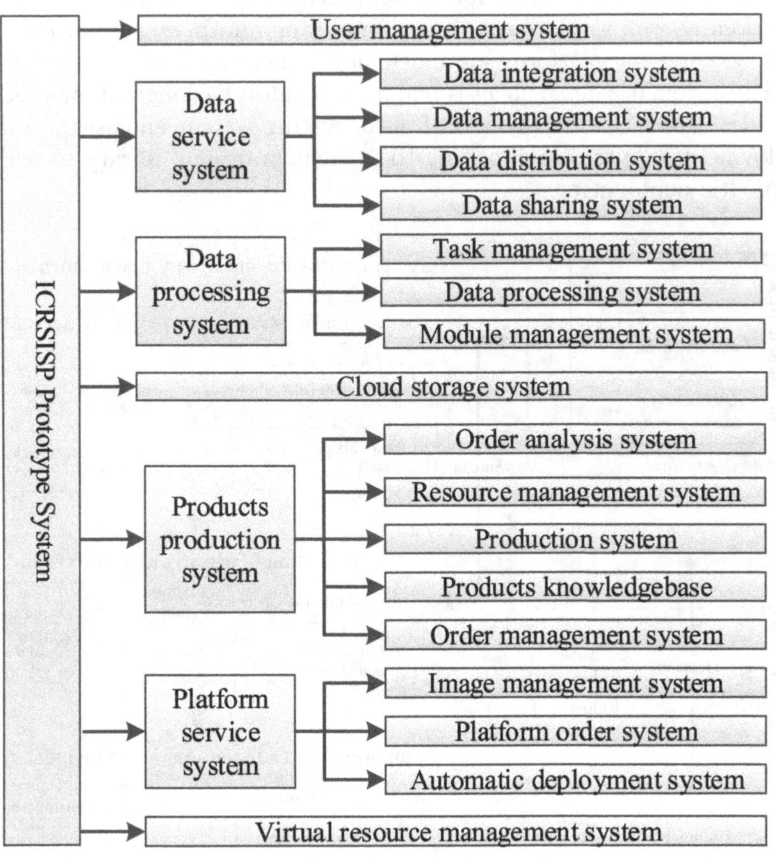

FIGURE 11.12: The system hierarchy of the CRSIS system.

The users of the CRSIS system are divided into two categories: administrators and ordinary users. Ordinary users can log into the CRSIS system for user registration, data retrieval and ordering, data downloading and sharing, data processing orders submission and obtaining processing results, product production orders submission and results acquisition, computing platform orders submission and results acquisition. In addition to all the rights of ordinary users, administrators can also perform distributed data integration and management, CRSIS system maintenance, etc. (Figure 11.13).

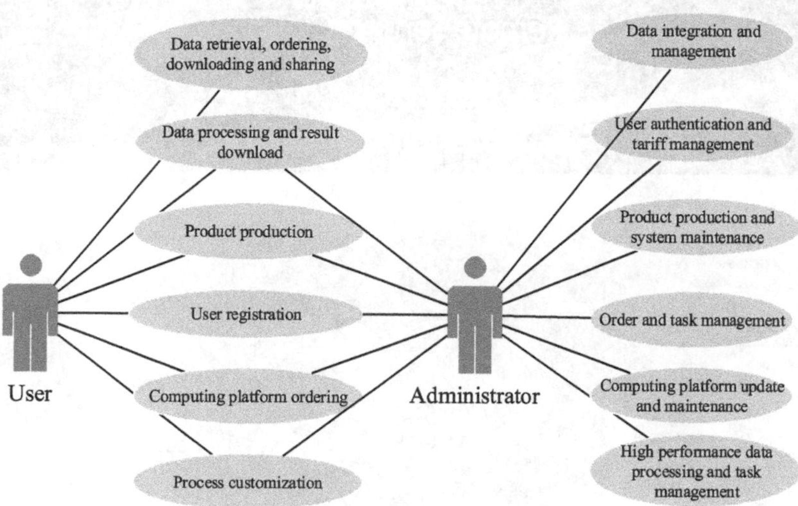

FIGURE 11.13: The user role of the CRSIS system.

The web interface development of the CRSIS system is based on the Spring MVC (Model View Controller) and Bootstrap3 framework. The welcome interface of the prototype system is shown in Figure 11.14.

FIGURE 11.14: The welcome page of the CRSIS system.

11.5.1 RSDaaS subsystem

The RSDaaS subsystem mainly provides functions such as retrieval, browsing, ordering, distribution sharing, and cloud storage of multi-source remote sensing data. Data retrieval includes two main types: single sensor retrieval and satellite network retrieval. The spatial range of data retrieval can be directly input into latitude and longitude information, or can be selected by mouse frame selection, or by administrative division (only in China); the time range can be continuous time interval or cross-year discontinuous time interval. The basic geographic base map of the data retrieval interface is published by the GeoServer, and the map interface operation is implemented by OpenLayers 3 [341] (Figure 11.15).

FIGURE 11.15: The data retrieval page of the CRSIS system.

Remote sensing data cloud storage is mainly implemented by OpenStack-Swift. Through applying for a cloud storage disk, users can obtain cloud storage services, including data uploading, downloading, sharing, directory creation and deletion, storage expansion and in-disk search(Figure 11.16).

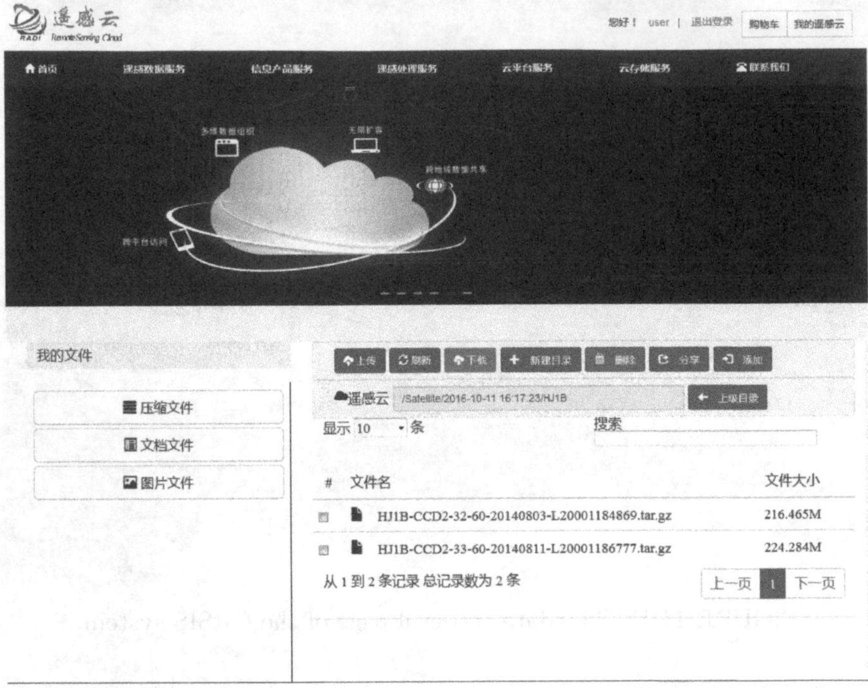

FIGURE 11.16: The personal cloud storage page of the CRSIS system.

11.5.2 RSDPaaS subsystem

The RSDPaaS subsystem mainly provides more than 90 kinds of serial and parallel remote sensing image processing algorithms, including level 0-2 remote sensing data preprocessing, remote sensing image enhancement, remote sensing data transformation, feature extraction, mosaic, fusion, etc. [328]. It can provide product production services for agriculture, forestry, minerals, and oceans applications (Figure 11.17).

FIGURE 11.17: The data processing page of the CRSIS system.

11.5.3 RSPPaaS subsystem

The RSPPaaS system mainly provides three types of products production services: fine processing products production, inversion index products production, and thematic products production [329, 342]. The fine processing mainly includes geometric normalization, radiation normalization, fusion, mosaic, etc. The inversion index products mainly include about 40 kinds of remote sensing inversion index products, including the Normalized Difference Vegetation Index(NDVI), Normalized Difference Water Index(NDWI), and so on[343, 344, 345]. Thematic products mainly include direct-application-oriented remote sensing data products or thematic maps, including agricultural topics, forestry topics, mineral topics, marine topics and other application fields. Using the RSPPaaS subsystem, users can submit product production orders by selecting the product type, time range, and spatial range. Through the production order analysis, workflow organization and execution, order status polling, products return and archiving, the production system finally feeds the product production results to the user for direct download or transfer to the personal cloud storage (Figure 11.18).

FIGURE 11.18: The remote sensing product production page of the CRSIS system.

11.5.4 RSCPaaS subsystem

The RSCPaaS subsystem mainly provides automatic deployment service for Windows and Linux operating systems, ENVI and Erdas Image softwares, GDAL and OpenCV libraries, etc. Without professional knowledge of the cloud, users also can obtain their wanted systems and the applications expediently [346](Figure 11.19).

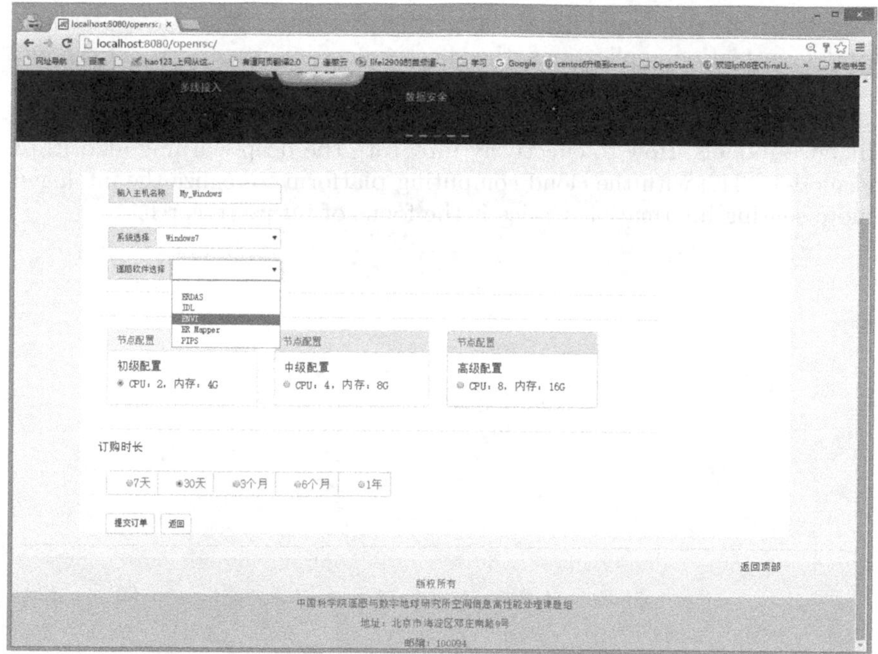

FIGURE 11.19: The remote sensing platform service page of the CRSIS system.

11.6 Conclusions

With the development of Earth Observation technology, the system architecture and computing model of the traditional remote sensing information service cannot meet the increasing user demand. Using cloud computing technology in the field of remote sensing can effectively improve the level and timeliness of remote sensing information services. Therefore, in this study, we designed and implemented the CRSIS system. The proposed CRSIS system provides users with RSDaaS, RSDPaaS, RSPPaaS and RSCPaaS, four types of user-oriented cloud services. Specifically, RSDaaS mainly includes distributed data integration, management, multi-user distribution and cloud storage sharing services; RSDPaaS contains Level 0-2 remote sensing data preprocessing, feature extraction, mosaic, fusion, etc.; RSPPaaS provides fine processing products, inversion index products and thematic products production services; RSCPaaS mainly provide users with remote-sensing-computing-oriented virtual machines. Finally, through the rapid retrieval of massive remote sensing image metadata, regional-level remote sensing image mosaic, global scale NPP products production, and virtual computing platform automatic deployment, four demonstration applications, the effectiveness and superiority of the integrated remote sensing information service cloud platform was verified.

However, this study did not consider the deep learning and artificial intelligence models. How to effectively integrate the deep learning algorithms and models [347] with the cloud computing platform to realize the intelligent remote sensing information service is the focus of future research.

Bibliography

[1] Russell Congalton. A review of assessing the accuracy of classifications of remotely sensed data. *Remote Sensing of Environment*, 37:35–46, 07 1991.

[2] Sabins, Jr. and Lulla, Kamlesh. Remote sensing: Principles and interpretation. *Geocarto International*, 2(1):66–66, 2009.

[3] Timothy I. Murphy. Remote sensing. `https://en.wikipedia.org/wiki/Remote_sensing`. Accessed April 4, 2010.

[4] Liu, Yuechen and Jiao, Weijie. Application of Remote Sensing Technology in Agriculture of the USA. *International Conference on Computer and Computing Technologies in Agriculture*, 107–114, Springer, 2015.

[5] D. Laney. 3d data management: Controlling data volume, velocity and variety. `https://blogs.gartner.com/doug-laney/files/2012/01/ad949-3D-Data-Management-Controlling-Data-Volume-Velocity-and-Variety.pdf`. Accessed February 6, 2001.

[6] Lizhe Wang, Jie Tao, Marcel Kunze, Alvaro Canales Castellanos, David Kramer, and Wolfgang Karl. Scientific cloud computing: Early definition and experience. In *High Performance Computing and Communications, 2008. HPCC'08. 10th IEEE International Conference on*, pages 825–830. IEEE, 2008.

[7] Lizhe Wang, Gregor Von Laszewski, Andrew Younge, Xi He, Marcel Kunze, Jie Tao, and Cheng Fu. Cloud computing: a perspective study. *New Generation Computing*, 28(2):137–146, 2010.

[8] Lizhe Wang, Marcel Kunze, Jie Tao, and Gregor von Laszewski. Towards building a cloud for scientific applications. *Advances in Engineering software*, 42(9):714–722, 2011.

[9] Peter Mell and Tim Grance. Nist definition of cloud computing. `https://csrc.nist.gov/publications/detail/sp/800-145/final`, 2011. Accessed September 3, 2018.

[10] Apache CloudStack Community. Apache cloudstack. `https://cloudstack.apache.org/`, 2017. Accessed September 4, 2018.

[11] OpenNebula.org. Opennebula project. `https://opennebula.org/`, 2008. Accessed September 28, 2018.

[12] Jeffrey Dean and Sanjay Ghemawat. Mapreduce: Simplified data processing on large clusters. In *6th Symposium on Operating System Design and Implementation (OSDI 2004), San Francisco, California, USA, December 6-8, 2004*, pages 137–150, 2004.

[13] Tom White. *Hadoop: The Definitive Guide*. O'Reilly Media, Inc., 4th edition, 2015.

[14] Konstantin Shvachko, Hairong Kuang, Sanjay Radia, and Robert Chansler. The hadoop distributed file system. In *Mass Storage Systems and Technologies (MSST), 2010 IEEE 26th Symposium on*, pages 1–10. Ieee, 2010.

[15] Javi Roman. The hadoop ecosystem table. `https://hadoopecosystemtable.github.io/`, 2014. Accessed September 28, 2018.

[16] Jens Dittrich and Jorge-Arnulfo Quiané-Ruiz. Efficient big data processing in hadoop mapreduce. *Proceedings of the VLDB Endowment*, 5(12):2014–2015, 2012.

[17] Matei Zaharia, Mosharaf Chowdhury, Tathagata Das, Ankur Dave, Justin Ma, Murphy McCauley, Michael J. Franklin, Scott Shenker, and Ion Stoica. Resilient distributed datasets: A fault-tolerant abstraction for in-memory cluster computing. In *Proceedings of the 9th USENIX conference on Networked Systems Design and Implementation*, pages 2–2. USENIX Association, 2012.

[18] Mike Franklin. The Berkeley data analytics stack: Present and future. In *Big Data, 2013 IEEE International Conference on*, pages 2–3. IEEE, 2013.

[19] Xiang Li and Lingling Wang. On the study of fusion techniques for bad geological remote sensing image. *J. Ambient Intelligence and Humanized Computing*, 6(1):141–149, 2015.

[20] Robert Jeansoulin. Review of forty years of technological changes in geomatics toward the big data paradigm. *ISPRS International Journal of Geo-Information*, 5(9):155, 2016.

[21] A. Lowe, D.; Mitchell. Status report on NASA's earth observing data and information system (eosdis). In *The 42nd Meeting of the Working Group on Information Systems & Services*, pages Frascati, Italy, 19–22 September, 2016.

[22] China's fy satellite data center. *Available online:* ***http://satellite.cma.gov.cn/portalsite/default.aspx***, (accessed on 25 August 2017).

[23] China center for resources satellite data and application. *Available online:* ***http://www.cresda.com/CN/sjfw/zxsj/index.shtml***, (accessed on 25 August 2017).

[24] Jining Yan and Lizhe Wang. Suitability evaluation for products generation from multisource remote sensing data. *Remote Sensing*, 8(12):995, 2016.

[25] Minggang Dou, Jingying Chen, Dan Chen, Xiaodao Chen, Ze Deng, Xuguang Zhang, Kai Xu, and Jian Wang. Modeling and simulation for natural disaster contingency planning driven by high-resolution remote sensing images. *Future Generation Computer Systems*, 37(C):367–377, 2014.

[26] Dongyao Wu, Liming Zhu, Xiwei Xu, Sherif Sakr, Daniel Sun, and Qinghua Lu. Building pipelines for heterogeneous execution environments for big data processing. *IEEE Software*, 33(2):60–67, 2016.

[27] Chaowei Yang, Qunying Huang, Zhenlong Li, Kai Liu, and Fei Hu. Big data and cloud computing: innovation opportunities and challenges. *International Journal of Digital Earth*, 10(1):13–53, 2017.

[28] Xin Luo, Maocai Wang, Guangming Dai, and Xiaoyu Chen. A novel technique to compute the revisit time of satellites and its application in remote sensing satellite optimization design. *International Journal of Aerospace Engineering,2017,(2017-01-31)*, 2017(6):1–9, 2017.

[29] Andrew Mitchell, Hampapuram Ramapriyan, and Dawn Lowe. Evolution of web services in eosdis — search and order metadata registry (echo). In *Geoscience and Remote Sensing Symposium,2009 IEEE International,igarss*, pages V–371 - V–374, 2010.

[30] Available online: `http://oodt.apache.org/`. Oodt. (accessed on 25 January 2017).

[31] Chris A. Mattmann, Daniel J. Crichton, Nenad Medvidovic, and Steve Hughes. A software architecture-based framework for highly distributed and data intensive scientific applications. In *International Conference on Software Engineering*, pages 721–730, 2006.

[32] Chris A. Mattmann, Dana Freeborn, Dan Crichton, Brian Foster, Andrew Hart, David Woollard, Sean Hardman, Paul Ramirez, Sean Kelly, and Albert Y. Chang. A reusable process control system framework for the orbiting carbon observatory and npp sounder peate missions. In *IEEE International Conference on Space Mission Challenges for Information Technology, 2009. Smc-It*, pages 165–172, 2009.

[33] Dennis C. Reuter, Cathleen M. Richardson, Fernando A. Pellerano, James R. Irons, Richard G. Allen, Martha Anderson, Murzy D. Jhabvala, Allen W. Lunsford, Matthew Montanaro, and Ramsey L. Smith. The thermal infrared sensor (tirs) on Landsat 8: Design overview and pre-launch characterization. *Remote Sensing*, 7(1):1135–1153, 2015.

[34] Khandelwal, Sumit and Goyal, Rohit. Effect of vegetation and urbanization over land surface temperature: case study of Jaipur City. *EARSeL Symposium*, pages 177–183, 2010.

[35] Yaxing Wei, Liping Di, Baohua Zhao, Guangxuan Liao, and Aijun Chen. Transformation of hdf-eos metadata from the ecs model to iso 19115-based xml. *Computers & Geosciences*, 33(2):238–247, 2007.

[36] Mahaxay, Manithaphone and Arunpraparut, Wanchai and Trisurat, Yongyut and Tangtham, Nipon. Modis: An alternative for updating land use and land cover in large river basin. *Thai J. For*, 33(3):34–47, 2014.

[37] Bo Zhong, Yuhuan Zhang, Tengteng Du, Aixia Yang, Wenbo Lv, and Qinhuo Liu. Cross-calibration of hj-1/ccd over a desert site using Landsat etm + imagery and Aster gdem product. *IEEE Transactions on Geoscience & Remote Sensing*, 52(11):7247–7263, 2014.

[38] Ranjeet Devarakonda, Giriprakash Palanisamy, Bruce E. Wilson, and James M. Green. Mercury: reusable metadata management, data discovery and access system. *Earth Science Informatics*, 3(1-2):87–94, 2010.

[39] Nengcheng Chen and Chuli Hu. A sharable and interoperable meta-model for atmospheric satellite sensors and observations. *IEEE Journal of Selected Topics in Applied Earth Observations & Remote Sensing*, 5(5):1519–1530, 2012.

[40] Peng Yue, Jianya Gong, and Liping Di. Augmenting geospatial data provenance through metadata tracking in geospatial service chaining. *Computers & Geosciences*, 36(3):270–281, 2010.

[41] Gilman, Jason Arthur and Shum, Dana. Making metadata better with cmr and mmt. NASA, 2016. `https://ntrs.nasa.gov/search.jsp?R=20160009277`

[42] Ann B. Burgess and Chris A. Mattmann. Automatically classifying and interpreting polar datasets with Apache tika. In *IEEE International Conference on Information Reuse and Integration*, pages 863–867, 2014.

[43] Chris Lynnes, Katie Baynes, and Mark McInerney. Big data in the earth observing system data and information system. Technical report, NASA Goddard Space Flight Center, June 2016.

[44] National Satellite Meteorological Centre. China's fy satellite data center. `http://satellite.nsmc.org.cn/portalsite/default.aspx`, 2013. Accessed February 28, 2018.

[45] Zhenyu Tan, Peng Yue, and Jianya Gong. An array database approach for earth observation data management and processing. *ISPRS International Journal of Geo-Information*, 6(7):220, 2017.

[46] Charles Toth and Grzegorz Jóźków. Remote sensing platforms and sensors: A survey. *ISPRS Journal of Photogrammetry and Remote Sensing*, 115:22–36, 2016.

[47] Mingmin Chi, Antonio Plaza, Jón Atli Benediktsson, Zhongyi Sun, Jinsheng Shen, and Yangyong Zhu. Big data for remote sensing: Challenges and opportunities. *Proceedings of the IEEE*, 104(11):2207–2219, 2016.

[48] Junqing Fan, Jining Yan, Yan Ma, and Lizhe Wang. Big data integration in remote sensing across a distributed metadata-based spatial infrastructure. *Remote Sensing*, 10(1):7, 2017.

[49] Xue Feng Lü, Cheng Qi Cheng, Jian Ya Gong, and Li Guan. Review of data storage and management technologies for massive remote sensing data. *Science China Technological Sciences*, 54(12):3220–3232, 2011.

[50] Khaled Nagi. Bringing search engines to the cloud using open source components. In *International Joint Conference on Knowledge Discovery, Knowledge Engineering and Knowledge Management*, pages 116–126, 2016.

[51] Jiyuan Li, Lingkui Meng, Frank Z. Wang, Wen Zhang, and Yang Cai. A map-reduce-enabled solap cube for large-scale remotely sensed data aggregation. *Computers & Geosciences*, 70:110–119, 2014.

[52] Ahmed Eldawy, Mohamed F. Mokbel, Saif Alharthi, Abdulhadi Alzaidy, Kareem Tarek, and Sohaib Ghani. Shahed: A mapreduce-based system for querying and visualizing spatio-temporal satellite data. In *Data Engineering (ICDE), 2015 IEEE 31st International Conference on*, pages 1585–1596. IEEE, 2015.

[53] M. Mazhar Rathore, Awais Ahmad, Anand Paul, and Alfred Daniel. Hadoop based real-time big data architecture for remote sensing earth observatory system. In *Computing, Communication and Networking Technologies (ICCCNT), 2015 6th International Conference on*, pages 1–7. IEEE, 2015.

[54] Sukanta Roy, Sanchit Gupta, and S.N. Omkar. Case study on: Scalability of preprocessing procedure of remote sensing in hadoop. *Procedia Computer Science*, 108:1672–1681, 2017.

[55] Yuehu Liu, Bin Chen, Wenxi He, and Yu Fang. Massive image data management using hbase and mapreduce. In *International Conference on Geoinformatics*, pages 1–5, 2013.

[56] Lin Wang, Chengqi Cheng, Shangzhu Wu, Feilong Wu, and Wan Teng. Massive remote sensing image data management based on hbase and geosot. In *Geoscience and Remote Sensing Symposium*, pages 4558–4561, 2015.

[57] Zhenwen He, Chonglong Wu, Gang Liu, Zufang Zheng, and Yiping Tian. Decomposition tree: a spatio-temporal indexing method for movement big data. *Cluster Computing*, 18(4):1481–1492, 2015.

[58] Wei Wang, Gilbert Cassar, Gilbert Cassar, and Klaus Moessner. An experimental study on geospatial indexing for sensor service discovery. *Expert Systems with Applications: An International Journal*, 42(7):3528–3538, 2014.

[59] Leptoukh, G. NASA remote sensing data in earth sciences: Processing, archiving, distribution, applications at the GES DISC. *Proc. of the 31st Intl Symposium of Remote Sensing of Environment*, 2005.

[60] Available online: `https://en.wikipedia.org/wiki/Geohash`. Geohash. (accessed on 25 August 2017).

[61] Zhe Yang, Weixin Zhai, Dong Chen, Wei Zhang, and Chengqi Cheng. A fast uav image stitching method on geosot. In *Geoscience and Remote Sensing Symposium*, pages 1785–1788, 2015.

[62] Dipayan Dev and Ripon Patgiri. Performance evaluation of hdfs in big data management. In *International Conference on High Performance Computing and Applications*, pages 1–7, 2015.

[63] Lars George. *HBase: the Definitive Guide: Random Access to your Planet-size Data.* O'Reilly Media, Inc., 2011.

[64] Xiaoming Gao, Vaibhav Nachankar, and Judy Qiu. Experimenting on lucene index on hbase in an hpc environment. In *Proceedings of the First Annual Workshop on High Performance Computing Meets Databases*, pages 25–28. ACM, 2011.

[65] Clinton Gormley and Zachary Tong. *Elasticsearch: The Definitive Guide: A Distributed Real-Time Search and Analytics Engine.* O'Reilly Media, Inc., 2015.

[66] Vishal Shukla. *Elasticsearch for Hadoop.* Packt Publishing Ltd, 2015.

[67] F. Pu, G. Cheng, C. Ren. Introduction for the subdivision and organization of spatial information. In *Science Press: Beijing, China*, 2012.

[68] Nan Lu, Chengqi Cheng, An Jin, and Haijian Ma. An index and retrieval method of spatial data based on geosot global discrete grid system. In *Geoscience and Remote Sensing Symposium*, pages 4519–4522, 2014.

[69] Jing Yan and Chengqi Cheng. Dynamic representation method of target in remote sensed images based on global subdivision grid. In *Geoscience and Remote Sensing Symposium*, pages 3097–3100, 2014.

[70] P.N. Happ, R.S. Ferreira, C.Bentes, G.A.O.P. Costa, and R.Q. Feitosa. Multiresolution segmentation: a parallel approach for high resolution image segmentation in multicore architectures. In *International Conference on Geographic Object-Based Image Analysis*, 2010.

[71] M. Vyverman, Baets B. De, V. Fack, and P. Dawyndt. Prospects and limitations of full-text index structures in genome analysis. *Nucleic Acids Research*, 40(15):6993–7015, 2012.

[72] Samneet Singh, Yan Liu, and Mehran Khan. Exploring cloud monitoring data using search cluster and semantic media wiki. In *IEEE Intl Conf on Ubiquitous Intelligence and Computing and 2015 IEEE Intl Conf on Autonomic and Trusted Computing and 2015 IEEE Intl Conf on Scalable Computing and Communications and ITS Associated Workshops*, pages 901–908, 2015.

[73] Available online: https://cwiki.apache.org/confluence/display/solr/Solr Cloud. Solrcloud. (accessed on 25 January 2017).

[74] Evie Kassela, Ioannis Konstantinou, and Nectarios Koziris. A generic architecture for scalable and highly available content serving applications in the cloud. In *IEEE Symposium on Network Cloud Computing and Applications*, pages 83–90, 2015.

[75] Jun Bai. Feasibility analysis of big log data real time search based on hbase and elasticsearch. In *Ninth International Conference on Natural Computation*, pages 1166–1170, 2014.

[76] Roberto Baldoni, Fabrizio Damore, Massimo Mecella, and Daniele Ucci. A software architecture for progressive scanning of on-line communities. In *IEEE International Conference on Distributed Computing Systems Workshops*, pages 207–212, 2014.

[77] Roberto Giachetta. A framework for processing large scale geospatial and remote sensing data in mapreduce environment. *Computers & Graphics*, 49:37–46, 2015.

[78] Yong Wang, Zhenling Liu, Hongyan Liao, and Chengjun Li. Improving the performance of GIS polygon overlay computation with mapreduce for spatial big data processing. *Cluster Computing*, 18(2):507–516, 2015.

[79] Muhammad Mazhar, Ullah Rathore, Anand Paul, Awais Ahmad, Bo Wei Chen, Bormin Huang, and Wen Ji. Real-time big data analytical architecture for remote sensing application. *IEEE Journal of Selected Topics in Applied Earth Observations & Remote Sensing*, 8(10):4610–4621, 2017.

[80] Nick Skytland. What is NASA doing with Big Data today?. *National Aeronautics and Space Administration*, 2012. `https://open.nasa.gov/blog/what-is-nasa-doing-with-big-data-today/`.

[81] Open Geospatial Consortium. The Opengis Abstract Specification Topic 7: The Earth Imagery Case. *Open Geospatial Consortium*, 2004. `https://portal.opengeospatial.org/files/?artifact_id=7467`.

[82] P. Gamba, Peijun Du, C. Juergens, and D. Maktav. Foreword to the special issue on "human settlements: A global remote sensing challenge". *Selected Topics in Applied Earth Observations and Remote Sensing, IEEE Journal of*, 4(1):5–7, March 2011.

[83] C.A. Lee, S.D. Gasster, A. Plaza, Chein-I Chang, and Bormin Huang. Recent developments in high performance computing for remote sensing: A review. *Selected Topics in Applied Earth Observations and Remote Sensing, IEEE Journal of*, 4(3):508–527, Sept 2011.

[84] A. Rosenqvist, M. Shimada, B. Chapman, A. Freeman, G. De Grandi, S. Saatchi, and Y. Rauste. The global rain forest mapping project-a review. *International Journal of Remote Sensing*, 21(6-7):1375–1387, 2000.

[85] Mathieu Fauvel, Jon Atli Benediktsson, John Boardman, John Brazile, Lorenzo Bruzzone, Gustavo Camps-Valls, Jocelyn Chanussot, Paolo Gamba, A. Gualtieri, M. Marconcini, et al. Recent advances in techniques for hyperspectral image processing. *Remote Sensing of Environment*, pages 1–45, 2007.

[86] Antonio J. Plaza. Special issue on architectures and techniques for real-time processing of remotely sensed images. *J. Real-Time Image Processing*, 4(3):191–193, 2009.

[87] Dongjian Xue, Zhengwei He, and Zhiheng Wang. Zhouqu county 8.8 extra-large-scale debris flow characters of remote sensing image analysis. In *Electronics, Communications and Control (ICECC), 2011 International Conference on*, pages 597–600, Sept 2011.

[88] G. Schumann, R. Hostache, C. Puech, L. Hoffmann, P. Matgen, F. Pappenberger, and L. Pfister. High-resolution 3-d flood information from radar imagery for flood hazard management. *Geoscience and Remote Sensing, IEEE Transactions on*, 45(6):1715–1725, June 2007.

[89] A.H.S. Solberg. Remote sensing of ocean oil-spill pollution. *Proceedings of the IEEE*, 100(10):2931–2945, Oct 2012.

[90] Bingxin Liu, Ying Li, Peng Chen, Yongyi Guan, and Junsong Han. Large oil spill surveillance with the use of modis and avhrr images. In *Remote Sensing, Environment and Transportation Engineering (RSETE), 2011 International Conference on*, pages 1317–1320, June 2011.

[91] A. Rosenqvist, M. Shimada, B. Chapman, K. McDonald, G. De Grandi, H. Jonsson, C. Williams, Y. Rauste, M. Nilsson, D. Sango, and M. Matsumoto. An overview of the jers-1 sar global boreal forest mapping (gbfm) project. In *Geoscience and Remote Sensing Symposium, 2004. IGARSS '04. Proceedings. 2004 IEEE International*, volume 2, pages 1033–1036 vol.2, Sept 2004.

[92] G. De Grandi, P. Mayaux, Y. Rauste, A. Rosenqvist, M. Simard, and S.S. Saatchi. The global rain forest mapping project jers-1 radar mosaic of tropical Africa: development and product characterization aspects. *Geoscience and Remote Sensing, IEEE Transactions on*, 38(5):2218–2233, Sep 2000.

[93] Antonio J. Plaza and Chein-I Chang. *High Performance Computing in Remote Sensing*. Chapman & Hall/CRC, 2007.

[94] Yan Ma, Lizhe Wang, Albert Y. Zomaya, Dan Chen, and Rajiv Ranjan. Task-tree based large-scale mosaicking for remote sensed imageries with dynamic dag scheduling. *IEEE Transactions on Parallel and Distributed Systems*, 99(PrePrints):1, 2013.

[95] Xinyuan Qu, Jiacun Li, Wenji Zhao, Xiaoli Zhao, and Cheng Yan. Research on critical techniques of disaster-oriented remote sensing quick mapping. In *Multimedia Technology (ICMT), 2010 International Conference on*, pages 1–4, Oct 2010.

[96] Yuehu Liu, Bin Chen, Hao Yu, Yong Zhao, Zhou Huang, and Yu Fang. Applying gpu and posix thread technologies in massive remote sensing image data processing. In *Geoinformatics, 2011 19th International Conference on*, pages 1–6, June 2011.

[97] Yan Ma, Lingjun Zhao, and Dingsheng Liu. An asynchronous parallelized and scalable image resampling algorithm with parallel i/o. In *Computational Science-ICCS 2009*, volume 5545 of *Lecture Notes in Computer Science*, pages 357–366. Springer Berlin Heidelberg, 2009.

[98] Min Cao and Zhao-liang Shi. Primary study of massive imaging autoprocessing system pixel factory. *Bulletin of Surveying and Mapping*, 10:55–58, 2006.

[99] Pandey, Suraj and Barker, Adam and Gupta, Kapil Kumar and Buyya, Rajkumar. Minimizing execution costs when using globally distributed cloud services. In *2010 24th IEEE International Conference on Advanced Information Networking and Applications*, pages 222–229, IEEE, 2010.

[100] Mandl, Daniel. Matsu: An Elastic Cloud Connected to a SensorWeb for Disaster Response. *National Aeronautics and Space Administration*, 2011. `https://ntrs.nasa.gov/archive/nasa/casi.ntrs.nasa.gov/20110008678.pdf`

[101] K. Keahey and M. Parashar. Enabling on-demand science via cloud computing. *Cloud Computing, IEEE*, 1(1):21–27, May 2014.

[102] Jun Xie, Yujie Su, Zhaowen Lin, Yan Ma, and Junxue Liang. Bare metal provisioning to openstack using xcat. *Journal of Computers(JCP)*, 8(7):1691–1695, 2013.

[103] A. Remon, S. Sanchez, A. Paz, E.S. Quintana-Orti, and A. Plaza. Real-time endmember extraction on multicore processors. *Geoscience and Remote Sensing Letters, IEEE*, 8(5):924–928, Sept 2011.

[104] R. Rabenseifner, G. Hager, and G. Jost. Hybrid mpi/openmp parallel programming on clusters of multi-core smp nodes. In *Parallel, Distributed and Network-based Processing, 2009 17th Euromicro International Conference on*, pages 427–436, Feb 2009.

[105] Plaza A., Qian Du, Yang-Lang Chang, and King R.L. High performance computing for hyperspectral remote sensing. *Selected Topics in Applied Earth Observations and Remote Sensing, IEEE Journal of*, 4(3):528–544, Sept 2011.

[106] Yinghui Zhao. Remote sensing based soil moisture estimation on high performance pc server. In *Environmental Science and Information Application Technology (ESIAT), 2010 International Conference on*, volume 1, pages 64–69, July 2010.

[107] Yanying Wang, Yan Ma, Peng Liu, Dingsheng Liu, and Jibo Xie. An optimized image mosaic algorithm with parallel io and dynamic grouped parallel strategy based on minimal spanning tree. In *Grid and Cooperative Computing (GCC), 2010 9th International Conference on*, pages 501–506, Nov 2010.

[108] Xue Xiaorong, Guo Lei, Wang Hongfu, and Xiang Fang. A parallel fusion method of remote sensing image based on ihs transformation. In *Image and Signal Processing (CISP), 2011 4th International Congress on*, volume 3, pages 1600–1603, Oct 2011.

[109] Taeyoung Kim, Myungjin Choi, and Tae-Byeong Chae. Parallel processing with mpi for inter-band registration in remote sensing. In *Parallel*

and Distributed Systems (ICPADS), 2011 IEEE 17th International Conference on, pages 1021–1025, Dec 2011.

[110] Yan Ma, Lizhe Wang, Dingsheng Liu, Peng Liu, Jun Wang, and Jie Tao. Generic parallel programming for massive remote sensing data processing. In *Cluster Computing (CLUSTER), 2012 IEEE International Conference on*, pages 420–428, Sept 2012.

[111] Filip Blagojevi, Paul Hargrove, Costin Iancu, and Katherine Yelick. Hybrid pgas runtime support for multicore nodes. In *Proceedings of the Fourth Conference on Partitioned Global Address Space Programming Model*, PGAS '10, pages 3:1–3:10, New York, NY, USA, 2010. ACM.

[112] Wei-Yu Chen, C. Iancu, and K. Yelick. Communication optimizations for fine-grained upc applications. In *Parallel Architectures and Compilation Techniques, 2005. PACT 2005. 14th International Conference on*, pages 267–278, Sept 2005.

[113] Nan Dun and K. Taura. An empirical performance study of chapel programming language. In *Parallel and Distributed Processing Symposium Workshops PhD Forum (IPDPSW), 2012 IEEE 26th International*, pages 497–506, May 2012.

[114] J. Milthorpe, V. Ganesh, AP. Rendell, and D. Grove. X10 as a parallel language for scientific computation: Practice and experience. In *Parallel Distributed Processing Symposium (IPDPS), 2011 IEEE International*, pages 1080–1088, May 2011.

[115] Ciprian Dobre and Fatos Xhafa. Parallel programming paradigms and frameworks in big data era. *International Journal of Parallel Programming*, 42(5):710–738, 2014.

[116] D. Nurmi, R. Wolski, C. Grzegorczyk, G. Obertelli, S. Soman, L. Youseff, and D. Zagorodnov. The eucalyptus open-source cloud-computing system. In *Cluster Computing and the Grid, 2009. CCGRID '09. 9th IEEE/ACM International Symposium on*, pages 124–131, May 2009.

[117] Jeffrey Dean and Sanjay Ghemawat. Mapreduce: Simplified data processing on large clusters. *Commun. ACM*, 51(1):107–113, January 2008.

[118] Feng-Cheng Lin, Lan-Kun Chung, Wen-Yuan Ku, Lin-Ru Chu, and Tien-Yin Chou. Service component architecture for geographic information system in cloud computing infrastructure. In *Advanced Information Networking and Applications (AINA), 2013 IEEE 27th International Conference on*, pages 368–373, March 2013.

[119] Bo Li, Hui Zhao, and Zhenhua Lv. Parallel isodata clustering of remote sensing images based on mapreduce. In *Cyber-Enabled Distributed Computing and Knowledge Discovery (CyberC), 2010 International Conference on*, pages 380–383, Oct 2010.

[120] Mohamed H. Almeer. Cloud hadoop map reduce for remote sensing image analysis. *Journal of Emerging Trends in Computing and Information Sciences*, 3(4):637–644, April 2012.

[121] R. Nasim and A.J. Kassler. Deploying openstack: Virtual infrastructure or dedicated hardware. In *Computer Software and Applications Conference Workshops (COMPSACW), 2014 IEEE 38th International*, pages 84–89, July 2014.

[122] A.B.M. Moniruzzaman, K.W. Nafi, and S.A. Hossain. An experimental study of load balancing of opennebula open-source cloud computing platform. In *Informatics, Electronics Vision (ICIEV), 2014 International Conference on*, pages 1–6, May 2014.

[123] Paul Barham, Boris Dragovic, Keir Fraser, Steven Hand, Tim Harris, Alex Ho, Rolf Neugebauer, Ian Pratt, and Andrew Warfield. Xen and the art of virtualization. In *Proceedings of the Nineteenth ACM Symposium on Operating Systems Principles*, SOSP '03, pages 164–177, New York, NY, USA, 2003. ACM.

[124] A. Kivity, Y. Kamay, D. Laor, U. Lublin, and A. Liguori. Kvm: the linux virtual machine monitor. In *OLS '09: Ottawa Linux Symposium 2009*, pages 225–230, Jul 2007.

[125] Qasim Ali, Vladimir Kiriansky, Josh Simons, and Puneet Zaroo. Performance evaluation of hpc benchmarks on vmware's esxi server. In Michael Alexander and P Da Ambra, editors, *Euro-Par 2011: Parallel Processing Workshops*, volume 7155 of *Lecture Notes in Computer Science*, pages 213–222. Springer Berlin Heidelberg, 2012.

[126] Yi-Man Ma, Che-Rung Lee, and Yeh-Ching Chung. Infiniband virtualization on kvm. In *Cloud Computing Technology and Science (CloudCom), 2012 IEEE 4th International Conference on*, pages 777–781, Dec 2012.

[127] S. Varrette, M. Guzek, V. Plugaru, X. Besseron, and P. Bouvry. Hpc performance and energy-efficiency of xen, kvm and vmware hypervisors. In *Computer Architecture and High Performance Computing (SBAC-PAD), 2013 25th International Symposium on*, pages 89–96, Oct 2013.

[128] S. Abdelwahab, B. Hamdaoui, M. Guizani, and A. Rayes. Enabling smart cloud services through remote sensing: An internet of everything enabler. *Internet of Things Journal, IEEE*, 1(3):276–288, June 2014.

[129] Mehul Nalin Vora. Hadoop-hbase for large-scale data. In *Computer Science and Network Technology (ICCSNT), 2011 International Conference on*, volume 1, pages 601–605, Dec 2011.

[130] Nan Lu, Chengqi Cheng, An Jin, and Haijian Ma. An index and retrieval method of spatial data based on geosot global discrete grid system.

In *Geoscience and Remote Sensing Symposium (IGARSS), 2013 IEEE International*, pages 4519–4522, July 2013.

[131] Jeremy Zawodny. Redis: Lightweight key/value store that goes the extra mile. *Linux Magazine*, 79, 2009.

[132] QI Jianghui, ZHANG Feng, DU Zhenhong, and LIU Renyi. Research of the landuse vector data storage and spatial index based on the main memory database. *Journal of Zhejiang University(Science Edition)*, 13(3):365–370, 2015.

[133] Chang, Victor and Kuo, Yen-Hung and Ramachandran, Muthu. Cloud computing adoption framework: A security framework for business clouds. *Future Generation Computer Systems*, 57:24–41, 2016, Elsevier.

[134] Meixia Deng, Liping Di, Genong Yu, A. Yagci, Chunming Peng, Bei Zhang, and Dayong Shen. Building an on-demand web service system for global agricultural drought monitoring and forecasting. In *Geoscience and Remote Sensing Symposium (IGARSS), 2012 IEEE International*, pages 958–961, July 2012.

[135] Cui Lin, Shiyong Lu, Xubo Fei, A. Chebotko, Darshan Pai, Zhaoqiang Lai, F. Fotouhi, and Jing Hua. A reference architecture for scientific workflow management systems and the view soa solution. *Services Computing, IEEE Transactions on*, 2(1):79–92, Jan 2009.

[136] Bertram Ludscher, Bertram, Ilkay Altintas, Chad Berkley, Dan Higgins, Efrat Jaeger, Matthew Jones, Edward A. Lee, Jing Tao, and Yang Zhao. Scientific workflow management and the Kepler system. *Concurrency and Computation: Practice and Experience*, 18(10):1039–1065, 2006.

[137] Ewa Deelman, Gurmeet Singh, Mei hui Su, James Blythe, Yolanda Gil, Carl Kesselman, Gaurang Mehta, Karan Vahi, G. Bruce Berriman, John Good, Anastasia Laity, Joseph C. Jacob, and Daniel S. Katz. Pegasus: a framework for mapping complex scientific workflows onto distributed systems. *Scientific Programming Journal*, 13:219–237, 2005.

[138] OpenMP Application Program Interface. https://www.openmp.org/. Accessed April 30, 2019.

[139] Kee, Yang-Suk. OpenMP extension to SMP clusters. *IEEE Potentials*, 25(3):37–42, IEEE, 2006.

[140] MPI: A Message-Passing Interface Standard Version 3.0. http://www.mpi-forum.org. (accessed on November 12, 2010).

[141] Rolf Rabenseifner, Georg Hager, and Gabriele Jost. Hybrid mpi/openmp parallel programming on clusters of multi-core smp nodes. In *Euromicro International Conference on Parallel, Distributed and Network-Based Processing*, pages 427–436, 2009.

[142] Murray Cole. Bringing skeletons out of the closet: a pragmatic manifesto for skeletal parallel programming. *Parallel Computing*, 30(3):389–406, 2004.

[143] M. Aldinucci, M. Danelutto, and P. Teti. An advanced environment supporting structured parallel programming in Java. *Future Generation Computer Systems*, 19(5):611–626, 2003.

[144] Philipp Ciechanowicz, Michael Poldner, and Herbert Kuchen. The münster skeleton library muesli: A comprehensive overview. *General Information*, (35):B360–B361, 2009.

[145] Jeffrey Dean and Sanjay Ghemawat. *MapReduce: simplified data processing on large clusters*. ACM, 2008.

[146] A. Stepanov and M. Lee. The standard template library. *Technical Report HPL-95-11, Hewlett-Packard Laboratories*, 1995.

[147] J. Falcou, J. Sérot, T. Chateau, and J. T. Lapresté. Quaff : efficient c++ design for parallel skeletons. *Parallel Computing*, 32(7):604–615, 2007.

[148] Yan Ma, Lingjun Zhao, and Dingsheng Liu. An asynchronous parallelized and scalable image resampling algorithm with parallel i/o. In *International Conference on Computational Science*, pages 357–366, 2009.

[149] Yanying Wang, Yan Ma, Peng Liu, Dingsheng Liu, and Jibo Xie. An optimized image mosaic algorithm with parallel io and dynamic grouped parallel strategy based on minimal spanning tree. In *International Conference on Grid and Cooperative Computing*, pages 501–506, 2011.

[150] Xinyuan Qu, Jiacun Li, Wenji Zhao, Xiaoli Zhao, and Cheng Yan. Research on critical techniques of disaster-oriented remote sensing quick mapping. In *International Conference on Multimedia Technology*, pages 1–4, 2010.

[151] Lingjun Zhao, Yan Ma, Guoqing Li, Wenyang Yu, and Jing Zhang. Rapid calculation research on water area extraction from asar image. In *Eighth International Conference on Grid and Cooperative Computing*, pages 339–343, 2009.

[152] Antonio J. Dios, Rafael Asenjo, Angeles Navarro, Francisco Corbera, and Emilio L. Zapata. High-level template for the task-based parallel wavefront pattern. In *International Conference on High Performance Computing*, 2011.

[153] Forrest G. Hall, Thomas Hilker, Nicholas C. Coops, Alexei Lyapustin, Karl F. Huemmrich, Elizabeth Middleton, Hank Margolis, Guillaume Drolet, and T. Andrew Black. Multi-angle remote sensing of forest light use efficiency by observing pri variation with canopy shadow fraction. *Remote Sensing of Environment*, 112(7):3201–3211, 2008.

[154] Ross S. Lunetta, Joseph F. Knight, Jayantha Ediriwickrema, John G. Lyon, and L. Dorsey Worthy. Land-cover change detection using multi-temporal modis ndvi data. *Remote Sensing of Environment*, 105(2):142–154, 2006.

[155] Matthew F. Mccabe and Eric F. Wood. Scale influences on the remote estimation of evapotranspiration using multiple satellite sensors. *Remote Sensing of Environment*, 105(4):271–285, 2006.

[156] NASA EOSDIS Web Site. http://www.esdis.eosdis.nasa.gov/. Accessed April 30, 2019.

[157] Mingmin Chi, Antonio Plaza, Jón Atli Benediktsson, Zhongyi Sun, Jinsheng Shen, and Yangyong Zhu. Big data for remote sensing: Challenges and opportunities. *Proceedings of the IEEE*, 104(11):2207–2219, 2016.

[158] Institute Of Remote Sensing Applications Chinese Academy Of Scences. http://english.radi.cas.cn/. Accessed April 30, 2019.

[159] Lizhe Wang, Ke Lu, Peng Liu, Rajiv Ranjan, and Lajiao Chen. Ik-svd: Dictionary learning for spatial big data via incremental atom update. *Computing in Science & Engineering*, 16(4):41–52, 2014.

[160] Lajiao Chen, Yan Ma, Peng Liu, Jingbo Wei, Wei Jie, and Jijun He. A review of parallel computing for large-scale remote sensing image mosaicking. *Cluster Computing*, 18(2):517–529, 2015.

[161] Lizhe Wang, Rajiv Ranjan, Joanna Kołodziej, Albert Zomaya, and Leila Alem. Software tools and techniques for big data computing in healthcare clouds. *Future Generation Computer Systems*, 43-44:38–39, 2015.

[162] Lizhe Wang, Dan Chen, Yangyang Hu, Yan Ma, and Jian Wang. Towards enabling cyberinfrastructure as a service in clouds. *Computers & Electrical Engineering*, 39(1):3–14, 2013.

[163] Dana Petcu, Dorian Gorgan, Textbfflorin Pop, Dacian Tudor, and Daniela Zaharie. Satellite image processing on a grid-based platform. *Advanced Engineering Materials*, 12(11):B618–B627, 2008.

[164] M. Esfandiari, H. Ramapriyan, J. Behnke, and E. Sofinowski. Earth observing system (eos) data and information system (eosdis) — evolution update and future. In *Geoscience and Remote Sensing Symposium, 2007. IGARSS 2007. IEEE International*, pages 4005–4008, 2007.

[165] Xiaolu Zhang, Jiafu Jiang, Xiaotong Zhang, and Xuan Wang. A data transmission algorithm for distributed computing system based on maximum flow. *Cluster Computing*, 18(3):1157–1169, 2015.

[166] S. Fiore, I. Epicoco, G. Quarta, G. Aloisio, M. Cafaro. Design and implementation of a grid computing environment for remote sensing. *High Performance Computing in Remote Sensing*, page 281, 2007.

[167] Asad Samar, Heinz Stockinger, Wolfgang Hoschek, Javier Jaen-Martinez and Kurt Stockinger. Data management in an international data grid project. *Lecture Notes in Computer Science*, 1971:77–90, 2000.

[168] Liping Di. The development of remote-sensing related standards at fgdc, ogc, and iso tc 211. In *Geoscience and Remote Sensing Symposium, 2003. IGARSS '03. Proceedings. 2003 IEEE International*, pages 643–647 vol.1, 2003.

[169] Adrian Colesa, Iosif Ignat, and Radu Opris. Providing high data availability in mediogrid. In *Eighth International Symposium on Symbolic and Numeric Algorithms for Scientific Computing*, pages 296–302, 2006.

[170] Osamu Tatebe, Kohei Hiraga, and Noriyuki Soda. Gfarm grid file system. *New Generation Computing*, 28(3):257–275, 2010.

[171] Chen Dan, Wang Lizhe, Holger, Ranjan Rajiv, and Chen Jingying. G-hadoop: Mapreduce across distributed data centers for data-intensive computing. *Future Generation Computer Systems*, 29(3):739–750, 2013.

[172] Kuang H.-Radia S., Chansler R., Shvachko, K. The hadoop distributed file system mass storage systems and technologies (msst) 2010. *Future Generation Computer Systems*, pages 1–10, 2010.

[173] Yong Wang, Zhenling Liu, Hongyan Liao, and Chengjun Li. Improving the performance of gis polygon overlay computation with mapreduce for spatial big data processing. *Cluster Computing*, 18(2):507–516, 2015.

[174] Ghoneimy, Adel and Zinatbaksh, Ali and Tiwari, Sandeep and Wein, Jerry D. and Jun, Andrew and Simhadri, Ruby S. Workflow system and method, 2008, july 22, Google Patents, US Patent 7,403,948.

[175] Guo, Huadong and Wang, Lizhe and Chen, Fang and Liang, Dong. Scientific big data and digital earth. *Chinese Science Bulletin*, 59(35):5066–5073, Springer, 2014.

[176] Jia Yu, Rajkumar Buyya, and Kotagiri Ramamohanarao. *Workflow Scheduling Algorithms for Grid Computing*. Springer Berlin Heidelberg, 2008.

[177] Lizhe Wang, Samee U. Khan, Dan Chen, Rajiv Ranjan, Cheng Zhong Xu, and Albert Zomaya. Energy-aware parallel task scheduling in a cluster. *Future Generation Computer Systems*, 29(7):1661–1670, 2013.

[178] Mihaela Catalina Nita, Florin Pop, Cristiana Voicu, Ciprian Dobre, and Fatos Xhafa. Momth: multi-objective scheduling algorithm of many tasks in hadoop. *Cluster Computing*, 18(3):1011–1024, 2015.

[179] Weijing Song, Shasha Yue, Lizhe Wang, Wanfeng Zhang, and Dingsheng Liu. Task scheduling of massive spatial data processing across distributed data centers: What's new? In *IEEE International Conference on Parallel and Distributed Systems*, pages 976–981, 2012.

[180] Wang L.-Liu D. Song W. Ma Y. Liu P. Chen D. Zhang, W. Towards building a multi-datacenter infrastructure for massive remote sensing image processing. *Concurrency and Computation: Practice and Experience*, 25(12):1798–1812, 2013.

[181] Ludäscher, Bertram and Altintas, Ilkay and Berkley, Chad and Higgins, Dan and Jaeger, Efrat and Jones, Matthew and Lee, Edward A. and Tao, Jing and Zhao, Yang. Scientific workflow management and the Kepler system. *Concurrency and Computation: Practice and Experience*, 18(10):1039–1065, Wiley Online Library, 2006.

[182] Efrat Jaeger, Ilkay Altintas, Jianting Zhang, Bertram Scher, Deana Pennington, and William Michener. A scientific workflow approach to distributed geospatial data processing using web services. In *International Conference on Scientific and Statistical Database Management, SSDBM 2005, 27-29 June 2005, University of California, Santa Barbara, Ca, Usa, Proceedings*, pages 87–90, 2005.

[183] Muthucumaru Maheswaran, Shoukat Ali, Howard Jay Siegel, Debra Hensgen, and Richard F. Freund. Dynamic matching and scheduling of a class of independent tasks onto heterogeneous computing systems. In *Heterogeneous Computing Workshop*, pages 30–44, 1999.

[184] Darren Quick and Kim-Kwang Raymond Choo. Impacts of increasing volume of digital forensic data: A survey and future research challenges. *Digital Investigation*, 11(4):273–294, 2014.

[185] Adam M. Wilson, Benoit Parmentier, and Walter Jetz. Systematic land cover bias in collection 5 modis cloud mask and derived products global overview. *Remote Sensing of Environment*, 141:149–154, 2014.

[186] M.C. Hansen, A. Egorov, P.V. Potapov, S.V. Stehman, A. Tyukavina, S.A. Turubanova, D.P. Roy, S.J. Goetz, T.R. Loveland, J. Ju, et al. Monitoring conterminous united states (conus) land cover change with web-enabled landsat data (weld). *Remote Sensing of Environment*, 140: 466–484, 2014.

[187] David P. Roy, Junchang Ju, KriSti Kline, Pasquale L. Scaramuzza, Valeriy Kovalskyy, Matthew Hansen, Thomas R. Loveland, Eric Vermote, and Chunsun Zhang. Web-enabled Landsat data (weld): Landsat etm+

composited mosaics of the conterminous united states. *Remote Sensing of Environment*, 114(1):35–49, 2010.

[188] Khondaker Abdullah Al Mamun, Musaed Alhussein, Kashfia Sailunaz, and Mohammad Saiful Islam. Cloud based framework for Parkinsons disease diagnosis and monitoring system for remote healthcare applications. *Future Generation Computer Systems*, 2015.

[189] Jayavardhana Gubbi, Rajkumar Buyya, Slaven Marusic, and Marimuthu Palaniswami. Internet of things (iot): A vision, architectural elements, and future directions. *Future Generation Computer Systems*, 29(7):1645–1660, 2013.

[190] Lizhe Wang, Marcel Kunze, Jie Tao, and Gregor von Laszewski. Towards building a cloud for scientific applications. *Advances in Engineering Software*, 42(9):714–722, 2011.

[191] Lizhe Wang and Cheng Fu. Research advances in modern cyberinfrastructure. *New Generation Computing*, 28(2):111–112, 2010.

[192] Lizhe Wang and Rajiv Ranjan. Processing distributed internet of things data in clouds. *IEEE Cloud Computing*, 2(1):76–80, 2015.

[193] Pengyao Wang, Jianqin Wang, Ying Chen, and Guangyuan Ni. Rapid processing of remote sensing images based on cloud computing. *Future Generation Computer Systems*, 29(8):1963–1968, 2013.

[194] Ze Deng, Yangyang Hu, Mao Zhu, Xiaohui Huang, and Bo Du. A scalable and fast OPTICS for clustering trajectory big data. *Cluster Computing*, 18(2):549–562, 2015.

[195] Yunliang Chen, Fangyuan Li, and Junqing Fan. Mining association rules in big data with NGEP. *Cluster Computing*, 18(2):577–585, 2015.

[196] Minggang Dou, Jingying Chen, Dan Chen, Xiaodao Chen, Ze Deng, Xuguang Zhang, Kai Xu, and Jian Wang. Modeling and simulation for natural disaster contingency planning driven by high-resolution remote sensing images. *Future Generation Computer Systems*, 37:367–377, 2014.

[197] Pengyao Wang, Jianqin Wang, Ying Chen, and Guangyuan Ni. Rapid processing of remote sensing images based on cloud computing. *Future Generation Computer Systems*, 29(8):1963–1968, 2013.

[198] Lizhe Wang, Tobias Kurze, Jie Tao, Marcel Kunze, and Gregor von Laszewski. On-demand service hosting on production grid infrastructures. *The Journal of Supercomputing*, 66(3):1178–1193, 2013.

[199] Zhenhua Lv, Yingjie Hu, Haidong Zhong, Jianping Wu, Bo Li, and Hui Zhao. Parallel k-means clustering of remote sensing images based on mapreduce. In *International Conference on Web Information Systems and Mining*, pages 162–170. Springer, 2010.

[200] Mohamed H. Almeer. Cloud hadoop map reduce for remote sensing image analysis. *Journal of Emerging Trends in Computing and Information Sciences*, 3(4):637–644, 2012.

[201] Feng-Cheng Lin, Lan-Kun Chung, Chun-Ju Wang, Wen-Yuan Ku, and Tien-Yin Chou. Storage and processing of massive remote sensing images using a novel cloud computing platform. *GIScience & Remote Sensing*, 50(3):322–336, 2013.

[202] Sofiane Bendoukha, Daniel Moldt, and Hayat Bendoukha. Building cloud-based scientific workflows made easy: A remote sensing application. In *International Conference of Design, User Experience, and Usability*, pages 277–288. Springer, 2015.

[203] Jianghao Wang, Yong Ge, Gerard B.M. Heuvelink, Chenghu Zhou, and Dick Brus. Effect of the sampling design of ground control points on the geometric correction of remotely sensed imagery. *International Journal of Applied Earth Observation and Geoinformation*, 18:91–100, 2012.

[204] Liangpei Zhang, Chen Wu, and Bo Du. Automatic radiometric normalization for multitemporal remote sensing imagery with iterative slow feature analysis. *IEEE Transactions on Geoscience and Remote Sensing*, 52(10):6141–6155, 2014.

[205] Jaewon Choi, Hyung-Sup Jung, and Sang-Ho Yun. An efficient mosaic algorithm considering seasonal variation: Application to kompsat-2 satellite images. *Sensors*, 15(3):5649–5665, 2015.

[206] Jixian Zhang. Multi-source remote sensing data fusion: status and trends. *International Journal of Image and Data Fusion*, 1(1):5–24, 2010.

[207] Baojuan Zheng, Soe W. Myint, Prasad S. Thenkabail, and Rimjhim M. Aggarwal. A support vector machine to identify irrigated crop types using time-series Landsat ndvi data. *International Journal of Applied Earth Observation and Geoinformation*, 34:103–112, 2015.

[208] Hao Sun, Xiang Zhao, Yunhao Chen, Adu Gong, and Jing Yang. A new agricultural drought monitoring index combining modis ndwi and day–night land surface temperatures: a case study in China. *International Journal of Remote Sensing*, 34(24):8986–9001, 2013.

[209] Anzhi Zhang and Gensuo Jia. Monitoring meteorological drought in semiarid regions using multi-sensor microwave remote sensing data. *Remote Sensing of Environment*, 134:12–23, 2013.

[210] Kun Xue, Yuchao Zhang, Hongtao Duan, Ronghua Ma, Steven Loiselle, and Minwei Zhang. A remote sensing approach to estimate vertical profile classes of phytoplankton in a eutrophic lake. *Remote Sensing*, 7(11):14403–14427, 2015.

[211] D.P. Roy, V. Kovalskyy, H.K. Zhang, E.F. Vermote, L. Yan, S.S. Kumar, and A. Egorov. Characterization of Landsat-7 to Landsat-8 reflective wavelength and normalized difference vegetation index continuity. *Remote Sensing of Environment*, 2016.

[212] L. Wang, J. Tao, Laszewski G. von, and H. Marten. Multicores in cloud computing: research challenges for applications. *Journal of Computers*, 5(6):958–964, 2010.

[213] Peng Liu, Tao Yuan, Yan Ma, Lizhe Wang, Dingsheng Liu, Shasha Yue, and Joanna Kolodziej. Parallel processing of massive remote sensing images in a GPU architecture. *Computing and Informatics*, 33(1):197–217, 2014.

[214] Changhe Song, Yunsong Li, and Bormin Huang. A gpu-accelerated wavelet decompression system with spiht and reed-solomon decoding for satellite images. *IEEE Journal of Selected Topics in Applied Earth Observations and Remote Sensing*, 4(3):683–690, 2011.

[215] Saurabh Kumar Garg, Steve Versteeg, and Rajkumar Buyya. A framework for ranking of cloud computing services. *Future Generation Computer Systems*, 29(4):1012–1023, 2013.

[216] Yingjie Xia, Xiumei Li, and Zhenyu Shan. Parallelized fusion on multisensor transportation data: a case study in cyberits. *International Journal of Intelligent Systems*, 28(6):540–564, 2013.

[217] Roberto Cossu, Elisabeth Schoepfer, Philippe Bally, and Luigi Fusco. Near real-time sar-based processing to support flood monitoring. *Journal of Real-Time Image Processing*, 4(3):205–218, 2009.

[218] Lionel Ménard, Isabelle Blanc, Didier Beloin-Saint-Pierre, Benoît Gschwind, Lucien Wald, Philippe Blanc, Thierry Ranchin, Roland Hischier, Simone Gianfranceschi, Steven Smolders, et al. Benefit of geoss interoperability in assessment of environmental impacts illustrated by the case of photovoltaic systems. *IEEE Journal of Selected Topics in Applied Earth Observations and Remote Sensing*, 5(6):1722–1728, 2012.

[219] Lizhe Wang, Yan Ma, Jining Yan, Victor Chang, and Albert Y. Zomaya. pipscloud: High performance cloud computing for remote sensing big data management and processing. *Future Generation Computer Systems*, 2016.

[220] Craig A. Lee, Samuel D. Gasster, Antonio Plaza, Chein-I Chang, and Bormin Huang. Recent developments in high performance computing for remote sensing: A review. *IEEE Journal of Selected Topics in Applied Earth Observations and Remote Sensing*, 4(3):508–527, 2011.

[221] Michael Armbrust, Armando Fox, Rean Griffith, Anthony D. Joseph, Randy Katz, Andy Konwinski, Gunho Lee, David Patterson, Ariel Rabkin, Ion Stoica, et al. A view of cloud computing. *Communications of the ACM*, 53(4):50–58, 2010.

[222] Hong Yao, Changmin Bai, Deze Zeng, Qingzhong Liang, and Yuanyuan Fan. Migrate or not? exploring virtual machine migration in roadside cloudlet-based vehicular cloud. *Concurrency and Computation: Practice and Experience*, 27(18):5780–5792, 2015.

[223] OpenStack. http://www.openstack.org/. Accessed April 30, 2019.

[224] Lizhe Wang, Yan Ma, Albert Y. Zomaya, Rajiv Ranjan, and Dan Chen. A parallel file system with application-aware data layout policies for massive remote sensing image processing in digital earth. *IEEE Transactions on Parallel and Distributed Systems*, 26(6):1497–1508, 2015.

[225] Yan Ma, Lizhe Wang, Peng Liu, and Rajiv Ranjan. Towards building a data-intensive index for big data computing–a case study of remote sensing data processing. *Information Sciences*, 319:171–188, 2015.

[226] Yan Ma, Lizhe Wang, Albert Y. Zomaya, Dan Chen, and Rajiv Ranjan. Task-tree based large-scale mosaicking for massive remote sensed imageries with dynamic dag scheduling. *IEEE Transactions on Parallel and Distributed Systems*, 25(8):2126–2137, 2014.

[227] Peng Yue, Liping Di, Yaxing Wei, and Weiguo Han. Intelligent services for discovery of complex geospatial features from remote sensing imagery. *ISPRS Journal of Photogrammetry and Remote Sensing*, 83:151–164, 2013.

[228] Wei Ren, Lingling Zeng, Ran Liu, and Chi Cheng. F2AC: A lightweight, fine-grained, and flexible access control scheme for file storage in mobile cloud computing. *Mobile Information Systems*, 2016:5232846:1–5232846:9, 2016.

[229] Wei Ren. uleepp: An ultra-lightweight energy-efficient and privacy-protected scheme for pervasive and mobile wbsn-cloud communications. *Ad Hoc & Sensor Wireless Networks*, 27(3-4):173–195, 2015.

[230] S. Afshin Mansouri, David Gallear, and Mohammad H. Askariazad. Decision support for build-to-order supply chain management through multiobjective optimization. *International Journal of Production Economics*, 135(1):24–36, 2012.

[231] Huadong Guo, Lizhe Wang, Fang Chen, et al. Scientific big data and digital earth. *Chinese Science Bulletin*, 59(35):5066–5073, 2014.

[232] Weijing Song, Lizhe Wang, Peng Liu, et al. Improved t-sne based manifold dimensional reduction for remote sensing data processing. *Multimedia Tools and Applications*, Feb 2018.

[233] Weijing Song, Lizhe Wang, Yang Xiang, et al. Geographic spatiotemporal big data correlation analysis via the Hilbert–Huang transformation. *Journal of Computer and System Sciences*, 89:130 – 141, 2017.

[234] Robert E. Kennedy, Zhiqiang Yang, and Warren B. Cohen. Detecting trends in forest disturbance and recovery using yearly Landsat time series: 1. landtrendr — temporal segmentation algorithms. *Remote Sensing of Environment*, 114(12):2897 – 2910, 2010.

[235] Toshihiro Sakamoto, Nhan Van Nguyen, Akihiko Kotera, et al. Detecting temporal changes in the extent of annual flooding within the Cambodia and the Vietnamese Mekong Delta from modis time-series imagery. *Remote Sensing of Environment*, 109(3):295 – 313, 2007.

[236] Zhang L.F., Chen H., Sun X.J., et al. Designing spatial-temporal-spectral integrated storage structure of multi-dimensional remote sensing images. *Journal of Remote Sensing*, 21(1):62 – 73, 2017.

[237] Assis, Luiz Fernando, Gilberto Ribeiro, et al. Big data streaming for remote sensing time series analytics using mapreduce. In *XVII Brazilian Symposium on GeoInformatics*, 2016.

[238] D.B. González and L.P. González. Spatial data warehouses and solap using open-source tools. In *2013 XXXIX Latin American Computing Conference (CLEI)*, pages 1–12, Oct 2013.

[239] T.O. Ahmed. Spatial on-line analytical processing (solap): Overview and current trends. In *2008 International Conference on Advanced Computer Theory and Engineering*, pages 1095–1099, Dec 2008.

[240] Lizhe Wang, Weijing Song, and Peng Liu. Link the remote sensing big data to the image features via wavelet transformation. *Cluster Computing*, 19(2):793–810, Jun 2016.

[241] Lizhe Wang, Jiabin Zhang, Peng Liu, et al. Spectral–spatial multi-feature-based deep learning for hyperspectral remote sensing image classification. *Soft Computing*, 21(1):213–221, Jan 2017.

[242] Weitao Chen, Xianju Li, Haixia He, et al. Assessing different feature sets' effects on land cover classification in complex surface-mined landscapes by Ziyuan-3 satellite imagery. *Remote Sensing*, 10(1), 2018.

[243] K.O. Asante, R.D. Macuacua, G.A. Artan, et al. Developing a flood monitoring system from remotely sensed data for the limpopo basin. *IEEE Transactions on Geoscience and Remote Sensing*, 45(6):1709–1714, June 2007.

[244] Matthew Rocklin. Dask: Parallel computation with blocked algorithms and task scheduling. In Kathryn Huff and James Bergstra, editors,

Proceedings of the 14th Python in Science Conference, pages 130 – 136, 2015.

[245] Erik Thomsen. *OLAP Solutions: Building Multidimensional Information Systems.* John Wiley Sons, Inc., 2002.

[246] Z. Yijiang. The conceptual design on spatial data cube. In *2012 2nd International Conference on Consumer Electronics, Communications and Networks (CECNet)*, pages 645–648, April 2012.

[247] Sonia Rivest, Yvan Bédard, Marie-Josée Proulx, Martin Nadeau, Frederic Hubert, and Julien Pastor. Solap technology: Merging business intelligence with geospatial technology for interactive spatio-temporal exploration and analysis of data. *ISPRS Journal of Photogrammetry and Remote Sensing*, 60(1):17 – 33, 2005.

[248] Matthew Scotch and Bambang Parmanto. Sovat: Spatial olap visualization and analysis tool. In *Proceedings of the Hawaii International Conference on System Sciences*, page 142.2, 2005.

[249] Junqing Fan, Jining Yan, Yan Ma, et al. Big data integration in remote sensing across a distributed metadata-based spatial infrastructure. *Remote Sensing*, 10(1), 2018.

[250] Jining Yan, Yan Ma, Lizhe Wang, et al. A cloud-based remote sensing data production system. *Future Generation Computer Systems*, 2017.

[251] Gilberto Camara, Luiz Fernando Assis, et al. Big earth observation data analytics: Matching requirements to system architectures. In *Proceedings of the 5th ACM SIGSPATIAL International Workshop on Analytics for Big Geospatial Data*, BigSpatial '16, pages 1–6, New York, NY, USA, 2016. ACM.

[252] SCIDB. A database management system designed for multidimensional data. `http://scidb.sourceforge.net/project.html`, 2017.

[253] Apache. Hadoop web site. `http://hadoop.apache.org/`, 2017.

[254] Konstantin Shvachko, Hairong Kuang, Sanjay Radia, et al. The hadoop distributed file system. In *IEEE Symposium on MASS Storage Systems and Technologies*, pages 1–10, 2010.

[255] odc. Open data cube. `http://datacube-core.readthedocs.io/en/latest/index.html`, 2017.

[256] OpenStreetMap. the project that creates and distributes free geographic data for the world. `http://www.openstreetmap.org`, 2017.

[257] UCAR. Netcdf file format and api. `http://www.unidata.ucar.edu/software/netcdf/`, 2017.

[258] Stephan Hoyer and Joseph J. Hamman. Xarray: N-d labeled arrays and datasets in python. *Journal of Open Research Software*, 5(3), 2017.

[259] Weitao Chen, Xianju Li, Haixia He, et al. A review of fine-scale land use and land cover classification in open-pit mining areas by remote sensing techniques. *Remote Sensing*, 10(1), 2018.

[260] Xianju Li, Weitao Chen, Xinwen Cheng, et al. Comparison and integration of feature reduction methods for land cover classification with rapideye imagery. *Multimedia Tools and Applications*, 76(21):23041–23057, Nov 2017.

[261] Xianju Li, Gang Chen, Jingyi Liu, et al. Effects of rapideye imagery's red-edge band and vegetation indices on land cover classification in an arid region. *Chinese Geographical Science*, 27(5):827–835, Oct 2017.

[262] Jie Zhang, Jining Yan, Yan Ma, et al. Infrastructures and services for remote sensing data production management across multiple satellite data centers. *Cluster Computing*, 19(3):1–18, 2016.

[263] Ye Tian, Xiong Li, Arun Kumar Sangaiah, et al. Privacy-preserving scheme in social participatory sensing based on secure multi-party cooperation. *Computer Communications*, 119:167 – 178, 2018.

[264] Chen Chen, Xiaomin Liu, Tie Qiu, et al. Latency estimation based on traffic density for video streaming in the internet of vehicles. *Computer Communications*, 111:176 – 186, 2017.

[265] W. Chen, X. Li, H. He, and L. Wang. Assessing different feature sets' effects on land cover classification in complex surface-mined landscapes by ziyuan-3 satellite imagery. *Remote Sensing*, 10(1), 2017.

[266] Jingbo Wei, Lizhe Wang, Peng Liu, Xiaodao Chen, Wei Li, and Albert Y. Zomaya. Spatiotemporal fusion of modis and Landsat-7 reflectance images via compressed sensing. *IEEE Transactions on Geoscience & Remote Sensing*, PP(99):1–14, 2017.

[267] Jining Yan, Yan Ma, Lizhe Wang, Kim Kwang Raymond Choo, and Wei Jie. A cloud-based remote sensing data production system. *Future Generation Computer Systems*, 2017.

[268] Yahya Aldhuraibi, Fawaz Paraiso, Nabil Djarallah, and Philippe Merle. Elasticity in cloud computing: State of the art and research challenges. *IEEE Transactions on Services Computing*, PP(99):1–1, 2018.

[269] Mohamed Abdel-Basset, Mohamed Mai, and Victor Chang. Nmcda: A framework for evaluating cloud computing services. *Future Generation Computer Systems*, 2018.

[270] Xiaobo Huang. Upgrading of enterprise internet technology architecture based on iaas and paas. *China Computer & Communication*, 2018.

[271] Syed Hamid Hussain Madni, Muhammad Shafie Abd Latiff, Yahaya Coulibaly, and Shafi'I Muhammad Abdulhamid. Resource scheduling for infrastructure as a service (iaas) in cloud computing: Challenges and opportunities. *Journal of Network & Computer Applications*, 68(C):173–200, 2016.

[272] Jin Li, Yinghui Zhang, Xiaofeng Chen, Yang Xiang, Jin Li, Yinghui Zhang, Xiaofeng Chen, and Yang Xiang. Secure attribute-based data sharing for resource-limited users in cloud computing. *Computers & Security*, 72, 2018.

[273] Victor Chang, Yen Hung Kuo, and Muthu Ramachandran. Cloud computing adoption framework. *Future Generation Computer Systems*, 57:24–41, 2016.

[274] R. Tanmay and Yogita Borse. Implementation of cloud computing service delivery models (iaas, paas) by aws and microsoft azure: A survey. *International Journal of Computer Applications*, 179(48):19–21, 2018.

[275] Yumin Wang, Jiangbo Li, and Harry Haoxiang Wang. Cluster and cloud computing framework for scientific metrology in flow control. *Cluster Computing*, (1):1–10, 2017.

[276] Jun Ye, Zheng Xu, and Yong Ding. *Secure Outsourcing of Modular Exponentiations in Cloud and Cluster Computing*. Kluwer Academic Publishers, 2016.

[277] Chao Shen, Weiqin Tong, Kim Kwang Raymond Choo, and Samina Kausar. Performance prediction of parallel computing models to analyze cloud-based big data applications. *Cluster Computing*, (1):1–16, 2017.

[278] Wei Guo, Jian Ya Gong, Wan Shou Jiang, Yi Liu, and Bing She. Openrs-cloud:a remote sensing image processing platform based on cloud computing environment. *Science China Technological Sciences*, 53(s1):221–230, 2010.

[279] Lizhe Wang, Peng Liu, Weijing Song, and Kim Kwang Raymond Choo. Duk-svd: dynamic dictionary updating for sparse representation of a long-time remote sensing image sequence. *Soft Computing*, (11):1–12, 2017.

[280] Jingbo Wei, Lizhe Wang, Peng Liu, and Weijing Song. Spatiotemporal fusion of remote sensing images with structural sparsity and semi-coupled dictionary learning. *Remote Sensing*, 9(1):21, 2016.

[281] Lin Heng Yang, Yao Yi Huang, Department Of Personne, and Quanzhou Normal University. Statistical information system design of Fujian province based on supermap iserver. *Geomatics & Spatial Information Technology*, 2017.

[282] Omar Sefraoui, Mohammed Aissaoui, and Mohsine Eleuldj. Openstack: Toward an open-source solution for cloud computing. *International Journal of Computer Applications*, 55(3):38–42, 2012.

[283] Lima, Stanley and Rocha, Alvaro. A view of OpenStack: toward an open-source solution for cloud. *World Conference on Information Systems and Technologies*, 481–491, Springer, 2017.

[284] Anton Beloglazov and Rajkumar Buyya. *OpenStack Neat: a Framework for Dynamic and Energy-efficient Consolidation of Virtual Machines in OpenStack Clouds.* John Wiley and Sons Ltd., 2015.

[285] Yongxia Jin and Ning Sun. Construction and application of cloud computing experimental platform based on openstack. *Experimental Technology & Management*, 2016.

[286] Rong Song. Design and research on automatic deployment and management system of cloud database based on saltstack. *Modern Information Technology*, 2017.

[287] Ziyi Wang and Chunhai Zhang. Design and implementation of a data visualization analysis component based on echarts. *Microcomputer & Its Applications*, 2016.

[288] L.I. Liang, Li Qin Tian, and L.I. Jun-Jian. Authentication and prediction of user behavior in cloud computing. *Computer Systems & Applications*, 2016.

[289] Jining Yan, Yan Ma, Lizhe Wang, Kim Kwang Raymond Choo, and Wei Jie. A cloud-based remote sensing data production system. *Future Generation Computer Systems*, 2017.

[290] Xiaoning Li, Lei Li, Lianwen Jin, and Desheng Li. Constructing a private cloud computing platform based on openstack. *Telecommunications Science*, 2012.

[291] Xianna Ji and Tsinghua University. To explore a behavior trust forecast game control mechanism based on trusted network. *Electronic Test*, 2015.

[292] Bingyu Zou, Huanguo Zhang, Xi Guo, H.U. Ying, and Jamila-Sattar. A fine-grained dynamic network authorization model based on trust level in trusted connecting network. *Journal of Wuhan University*, 56(2):147–150, 2010.

[293] X.U. Jian, L.I. Ming-Jie, Fu Cai Zhou, Rui Xue, and Northeastern University. Identity authentication method based on user's mouse behavior. *Computer Science*, 2016.

[294] Y.R. Chen, L.Q. Tian, and Y. Yang. Modeling and analysis of dynamic user behavior authentication scheme in cloud computing. *Journal of System Simulation*, 23(11):2302–2307, 2011.

[295] Jie Zhao. Behaviour trust control based on Bayesian networks and user behavior log mining. *Journal of South China University of Technology*, 37(5):94–100, 2009.

[296] Hongan Xie, L.I. Dong, S.U. Yang, and Kai Yang. Trusted network management model based on clustering analysis. *Journal of Computer Applications*, pages 220–226, 2016.

[297] Xiaoju Wang, Liqin Tian, Jingxiong Zhao, and Computer Department. User behavioral authentication mechanism and analysis based on iot. *Journal of Nanjing University of Science & Technology*, 39(1):70–77, 2015.

[298] Xiao Yi Yu and Aiming Wang. Steganalysis based on Bayesion network and genetic algorithm. In *International Congress on Image and Signal Processing*, pages 1–4, 2009.

[299] Zhiwei He, Mingyu Gao, Guojin Ma, Yuanyuan Liu, and Sanxin Chen. Online state-of-health estimation of lithium-ion batteries using dynamic Bayesian networks. *Journal of Power Sources*, 267(3):576–583, 2014.

[300] Grant McKenzie, Zheng Liu, Yingjie Hu, and Myeong Lee. Identifying urban neighborhood names through user-contributed online property listings. *ISPRS International Journal of Geo-Information*, 7(10), 2018.

[301] Ngo Manh Khoi and Sven Casteleyn. Analyzing spatial and temporal user behavior in participatory sensing. *ISPRS International Journal of Geo-Information*, 7(9), 2018.

[302] Jingbo Wei, Lizhe Wang, Peng Liu, Xiaodao Chen, Wei Li, and Albert Y. Zomaya. Spatiotemporal fusion of modis and Landsat-7 reflectance images via compressed sensing. *IEEE Transactions on Geoscience & Remote Sensing*, PP(99):1–14, 2017.

[303] William A. Branch and Bruce Mcgough. Replicator dynamics in a cobweb model with rationally heterogeneous expectations. *Journal of Economic Behavior & Organization*, 65(2):224–244, 2008.

[304] Nam N. Vo and Aaron F. Bobick. Sequential interval network for parsing complex structured activity. *Computer Vision and Image Understanding*, 143:147–158, 2016.

[305] Sudhansu Sekhar Nishank, Manoranjan Ranjit, Shantanu K. Kar, and Guru Prasad Chhotray. A model for analyzing the roles of network and user behavior in congestion control. *Physical Review C*, 49(1):320–323, 2004.

[306] Markus Wurzenberger, Florian Skopik, Giuseppe Settanni, and Wolfgang Scherrer. Complex log file synthesis for rapid sandbox-benchmarking of security- and computer network analysis tools. *Information Systems*, 60(C):13–33, 2016.

[307] Ines Brosso, Alessandro La Neve, Graça Bressan, and Wilson Vicente Ruggiero. A continuous authentication system based on user behavior analysis. In *Ares '10 International Conference on Availability, Reliability, and Security*, pages 380–385, 2010.

[308] L.I. Deren, Liangpei Zhang, and Guisong Xia. Automatic analysis and mining of remote sensing big data. *Acta Geodaetica Et Cartographica Sinica*, 2014.

[309] F. Lin, L. Chung, W. Ku, L. Chu, and T. Chou. Service component architecture for geographic information system in cloud computing infrastructure. In *2013 IEEE 27th International Conference on Advanced Information Networking and Applications (AINA)*, pages 368–373, March 2013.

[310] Fuhu Ren and Jinnian Wang. Turning remote sensing to cloud services: Technical research and experiment. *Journal of Remote Sensing*, (6):1331–1346, 2012.

[311] German Geoss Implementation Plan and D-GIP. *Global Earth Observation System of Systems (GEOSS)*. Springer New York, 2014.

[312] P. Smits and D. Rizzi. Activities in the European Commission: Inspire, gmes, and fp6. In *Theory & Applications of Knowledge-driven Image Information Mining with Focus on Earth Observation*, 2004.

[313] S.D. Peckham, C. Deluca, D.J. Gochis, J. Arrigo, A. Kelbert, E. Choi, and R. Dunlap. Earthcube - earth system bridge: Spanning scientific communities with interoperable modeling frameworks. In *AGU Fall Meeting*, 2014.

[314] Junxiao Hu, Jian Jiao, Haizhen Zhang, Xin Li, and Qiming Zeng. A new user-oriented remote sensing information service model based on integrated platform. In *Geoscience & Remote Sensing Symposium*, pages 5199–5202, 2014.

[315] W. Chen, X. Li, H. He, and L. Wang. A review of fine-scale land use and land cover classification in open-pit mining areas by remote sensing techniques. *Remote Sensing*, 10(1):15, 2018.

[316] Rajkumar Buyya, Christian Vecchiola, and S. Thamarai Selvi. *Mastering Cloud Computing: Foundations and Applications Programming.* Morgan Kaufmann Publishers Inc., 2013.

[317] Kobusińska, Anna and Leung, Carson and Hsu, Ching-Hsien and Raghavendra, S. and Chang, Victor. Emerging trends, issues and challenges in Internet of Things, Big Data and cloud computing. *Future Generation Computer Systems*, 87:416–419, 2018, https://doi.org/10.1016/j.future.2018.05.021.

[318] Luis M. Vaquero, Luis Rodero-Merino, Maik Lindner, and Maik Lindner. A break in the clouds: towards a cloud definition. *Acm Sigcomm Computer Communication Review*, 39(1):50–55, 2008.

[319] Juhnyoung Lee. A view of cloud computing. *Communications of the Acm*, 53(4):50–58, 2013.

[320] Lizhe Wang, Gregor Von Laszewski, Andrew Younge, Xi He, Marcel Kunze, Jie Tao, and Cheng Fu. Cloud computing: a perspective study. *New Generation Computing*, 28(2):137–146, 2010.

[321] Chaowei Yang, Michael Goodchild, Qunying Huang, Doug Nebert, Robert Raskin, Yan Xu, Myra Bambacus, and Daniel Fay. Spatial cloud computing: how can the geospatial sciences use and help shape cloud computing? *International Journal of Digital Earth*, 4(4):305–329, 2011.

[322] F.G. de Assis, Luiz and E.A. Horita, Flávio and P. de Freitas, Edison and Ueyama, Jó and de Albuquerque, João. A service-oriented middleware for integrated management of crowdsourced and sensor data streams in disaster management. *Sensors*, 18(6):1689, 2018, Multidisciplinary Digital Publishing Institute.

[323] X.U. Sheng, Zhi Chun Niu, and Zhen Bing Qian. Remote sensing monitoring of ecological environment in Anhui province during the 12 (th) five-year period. *Environmental Monitoring & Forewarning*, 2016.

[324] Dengrong Zhang, X.U. Siying, Bin Xie, W.U. Wenyuan, and L.U. Haifeng. Land use change of reclaimed mud flats in Jiaojiang- Taizhou Estuary in the past 40 years based on remote sensing technology. *Remote Sensing for Land & Resources*, 2016.

[325] A. Ismail, B.A. Bagula, and E. Tuyishimire. Internet-of-things in motion: A uav coalition model for remote sensing in smart cities. *Sensors*, 18(7), 2018.

[326] Francesco Vuolo, Mateusz Żółtak, Claudia Pipitone, Luca Zappa, Hannah Wenng, Markus Immitzer, Marie Weiss, Frederic Baret, and Clement Atzberger. Data service platform for sentinel-2 surface reflectance and value-added products: System use and examples. *Remote Sensing*, 8(11):938, 2016.

[327] Junqing Fan, Jining Yan, Yan Ma, and Lizhe Wang. Big data integration in remote sensing across a distributed metadata-based spatial infrastructure. *Remote Sensing*, 10(1):7, 2017.

[328] Lizhe Wang, Yan Ma, Jining Yan, Victor Chang, and Albert Y. Zomaya. pipscloud: High performance cloud computing for remote sensing big data management and processing. *Future Generation Computer Systems*, 2016.

[329] Jining Yan, Yan Ma, Lizhe Wang, Kim Kwang Raymond Choo, and Wei Jie. A cloud-based remote sensing data production system. *Future Generation Computer Systems*, 2017.

[330] Lin, Feng-Cheng and Chung, Lan-Kun and Ku, Wen-Yuan and Chu, Lin-Ru and Chou, Tien-Yin. The framework of cloud computing platform for massive remote sensing images. *2013 IEEE 27th International Conference on Advanced Information Networking and Applications (Aina)*, 621–628, 2013, IEEE.

[331] Abdel-Basset, Mohamed and Mohamed, Mai and Chang, Victor. NM-CDA: A framework for evaluating cloud computing services. *Future Generation Computer Systems*, 86:12–29, Elsevier, 2018.

[332] Hong Yao, Changmin Bai, Deze Zeng, Qingzhong Liang, and Yuanyuan Fan. Migrate or not? exploring virtual machine migration in roadside cloudlet-based vehicular cloud. *Concurrency & Computation Practice & Experience*, 27(18):5780–5792, 2016.

[333] Roberto Giachetta. A framework for processing large scale geospatial and remote sensing data in mapreduce environment. *Computers & Graphics*, 49(C):37–46, 2015.

[334] Mingruo Shi and Ruiping Yuan. Mad: A monitor system for big data applications. 2015.

[335] Victor Andres Ayma Quirita, Patrick Nigri Happ, Raul Queiroz Feitosa, Rodrigo Da Silva Ferreira, Dário Augusto Borges Oliveira, and Antonio Plaza. A new cloud computing architecture for the classification of remote sensing data. *IEEE Journal of Selected Topics in Applied Earth Observations & Remote Sensing*, PP(99):1–8, 2017.

[336] Gregory Giuliani, Bruno Chatenoux, Andrea De Bono, Denisa Rodila, Jean Philippe Richard, Karin Allenbach, Hy Dao, and Pascal Peduzzi. Building an earth observations data cube: lessons learned from the swiss data cube (sdc) on generating analysis ready data (ard). 1(1):1–18, 2017.

[337] Kun Jia, Linqing Yang, Shunlin Liang, Zhiqiang Xiao, Xiang Zhao, Yunjun Yao, Xiaotong Zhang, Bo Jiang, and Duanyang Liu. Long-term global land surface satellite (glass) fractional vegetation cover product

derived from modis and avhrr data. *IEEE Journal of Selected Topics in Applied Earth Observations & Remote Sensing*, PP(99).

[338] Lizhe Wang, Peng Liu, Weijing Song, and Kim Kwang Raymond Choo. Duk-svd: dynamic dictionary updating for sparse representation of a long-time remote sensing image sequence. *Soft Computing*, (11):1–12, 2017.

[339] Jingbo Wei, Lizhe Wang, Peng Liu, and Weijing Song. Spatiotemporal fusion of remote sensing images with structural sparsity and semi-coupled dictionary learning. *Remote Sensing*, 9(1):21, 2016.

[340] Chris A. Mattmann, Daniel J. Crichton, Nenad Medvidovic, and Steve Hughes. A software architecture-based framework for highly distributed and data intensive scientific applications. In *Proceedings of the 28th International Conference on Software Engineering*, ICSE '06, pages 721–730, 2006.

[341] Ying Lei, Ming Wang, Ying Qian Zhang, and Jing Jing Wang. Research and application of dynamic ogc service management technology based on geoserver. *Geomatics & Spatial Information Technology*, 2016.

[342] W. Chen, X. Li, H. He, and L. Wang. Assessing different feature sets' effects on land cover classification in complex surface-mined landscapes by Ziyuan-3 satellite imagery. *Remote Sensing*, 10(1), 2017.

[343] L. Wang, Y. Ma, A.Y. Zomaya, and R. Ranjan. A parallel file system with application-aware data layout policies for massive remote sensing image processing in digital earth. *IEEE Transactions on Parallel & Distributed Systems*, 26(6):1497–1508, 2015.

[344] Yan Ma, Lizhe Wang, Peng Liu, and Rajiv Ranjan. Towards building a data-intensive index for big data computing - a case study of remote sensing data processing. *Information Sciences*, 319(C):171–188, 2015.

[345] Yan Ma, Lizhe Wang, Albert Y. Zomaya, Dan Chen, and Rajiv Ranjan. Task-tree based large-scale mosaicking for massive remote sensed imageries with dynamic dag scheduling. *IEEE Transactions on Parallel & Distributed Systems*, 25(8):2126–2137, 2014.

[346] Rongqi Zhang, Yanlei Shang, and Si Zhang. An automatic deployment mechanism on cloud computing platform. In *IEEE International Conference on Cloud Computing Technology and Science*, pages 511–518, 2014.

[347] C. Liu, L. He, Z. Li, and J. Li. Feature-driven active learning for hyperspectral image classification. *IEEE Transactions on Geoscience and Remote Sensing*, 56(1):341–354, Jan 2018.

Index